RESIDENTIAL STEEL FRAMING HANDBOOK

RESIDENTIAL STEEL FRAMING HANDBOOK

Robert Scharff

and the

Editors of *Walls & Ceilings Magazine*

Boston, Massachusetts   Burr Ridge, Illinois
Dubuque, Iowa   Madison, Wisconsin   New York, New York
San Francisco, California   St. Louis, Missouri

McGraw-Hill

A Division of The McGraw-Hill Companies

Library of Congress Cataloging-in-Publication Data

Scharff, Robert
    Residential steel framing handbook / by Robert Scharff and the
editors of Walls & ceilings magazine.
        p.   cm.
    Includes index.
    ISBN 0-07-057231-3 (hc)
    1. Steel houses—Design and construction. 2. Framing (Building)
I. Walls & ceilings.   II. Title
TH4814.S73S35   1996
691'.8—dc20   95-52978
                                                        CIP

                8   9   BKM BKM   0 9 8 7 6 5 4 3

ISBN 0-07-057231-3

The sponsoring editor of this book was April D. Nolan, the book editor was Barbara B. Minich, and the production supervisor was Katherine G. Brown. This book was set in Times Roman. It was composed in Blue Ridge Summit, Pa.

McGraw-Hill books are available at special quantity discounts to use as premiums and sales promotions, or for use in corporate training programs. For more information, please write to the Director of Special Sales, McGraw-Hill, 11 West 19th Street, New York, NY 10011. Or contact your local bookstore.

Product or brand names used in this book may be trade names or trademarks. Where we believe that there may be proprietary claims to such trade names or trademarks, the name has been used with an initial capital or it has been capitalized in the style used by the name claimant. Regardless of the capitalization used, all such names have been used in an editorial manner without any intent to convey endorsement of or other affiliation with the name claimant. Neither the author nor the publisher intends to express any judgment as to the validity or legal status of any such proprietary claims.

*In order to receive additional information on these or any other McGraw-Hill
titles, in the United States please call 1-800-822-8158.*   0572313
In other countries, contact your local McGraw-Hill representative.   HB1A

CON

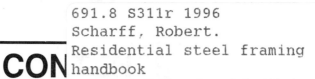

Chapter 5. Steel Floor Framing   117

Chapter 6. Load- and Nonload-Bearing Walls   159

Chapter 7. Steel-Framed Roofs   215

Chapter 8. Interior Framing Details   263

ACKNOWLEDGMENTS

I would like to thank the following people, corporations, and associates for their help in furnishing materials and illustrations used in this book.

Advanced Framing Systems, Inc.
Alabama Metal Industries Corp.
American Iron and Steel Institute
American Plywood Association
American Studco, Inc.
Angelist Metal Systems, Inc.
California Building System, Inc.
Clark Steel Framing Systems, Inc.
Consolidated Systems, Inc.
Dale/Incor Industries
Innovation Steel System
Interior Building Systems, Inc.
Knorr Steel Framing Systems
Light Gauge Steel Engineers Association
Madray Steel Buildings Systems
Marino Ware, Inc.
Metal Home Digest
Metal Stud Manufacturer's Association
National Gypsum Company
Quik Drive Fastening System, Inc.
ResSteel Framing System
Simpson Strong-Tie Company, Inc.
Steeler Co., Inc.
Steelman's Catalog, Inc.
Tri-Steel Structures, Inc.
John Brown of Tri-Steel Structures, Inc. for the use of the materials that appear in the
   Appendices.
Unimast Incorporated
United Construction Supply
United States Gypsum Company
Walls & Ceilings Magazine
Western Metal Lath

In addition, I would like to thank the following members of the Scharff Corporation for their help with editing and coordinating the text and art for this book: Leon E. Korejwo and Pamela L. Korejwo.

LIST OF ACRONYMS

The following symbols commonly are found in manufacturers' catalogs.

DEFINITION OF MANUFACTURERS' CATALOG SYMBOLS

A Cross-sectional area (IN^2)

AD Allowable deflection (IN)

ADEAD Collateral dead load

AE Effective cross-sectional area of the member used for axial load (IN^2)

Ae Effective area at stress Fn=Fy

AET Effective cross-sectional area based on stub column test

ANET Full cross section

ARE Allowable reaction exterior support (KIPS)

ARI Allowable reaction interior support (KIPS)

BAY Distance between supports (FT)

Beta Torsional-flexural constant

Cw Torsional-warping constant of the cross section

D Distance from Y-axis to outside of web

DEFL Value for deflection from load tables (IN)

DR Deflection requirement

ER Exterior reaction from calculations (KIPS)

EXT Exterior reaction from continuous-load tables (KIPS)

F Fixed factor used in computing allowable web crippling (KIPS)

Fu Tensile strength (KSI)

Fy Yield point (KSI)

h/t Ratio of depth

INT Interior reaction from continuous-load tables (KIPS)

I_X Moment of inertia about X-axis (IN^4)

I_Y Moment of inertia about Y-axis (IN^4)

I_{XW} Effective moment of inertia for deflection calculations (IN^4)

j   Sectional property for torsional-flexural buckling

J   St. venant torsion constant

KIP   1000 pounds

KL_t Torsional buckling length (FT)

KL_x Major axis support distance (FT)

KL_y Lateral support distance (FT)

Ksi   KIPS per square inch

L   Lip length (IN)

LE   Bearing length at end support (IN)

LIL   Bearing length at interior support for continuous (IN)

LIS   Bearing length at interior support for simple span (IN)

M_2 Allowable bending moment about X-X axis

Ma or Mr   Allowable bending moment (K-FT)

OC   Safety factor for axial compression

Pa   Allowable axial load from axial-load table (KIPS)

psi   Pounds per square inch

psf   Pounds per square foot

R_X Radius of gyration about X-axis (IN)

R_Y Radius of gyration about Y-axis (IN)

RCOEF   Roof-wind coefficient

REACT   Reaction from simple span load tables (KIPS)

Rext   Exterior reaction based on 1¼-inch supports (KIPS)

Rint   Interior reaction based on 1¼-inch supports (KIPS)

R.M.   Allowable resisting moment in KIP-inch.

Ro   Polar radius of gyration of cross section about the shear center

S_X Sectional modulus about the principal X-axis

S_Y Full section modulus about Y-axis (IN^3)

SE   Effective section modulus about X-axis (IN^3)

SPACE   Distance between adjacent purlins or girts (FT)

TRIB   Area supported by column (FT^2)

V Variable factor used in computing allowable web crippling

Va Allowable shear force (KIPS)

WCOEF Wall-wind coefficient

WT Section weight (LB/FT)

XB Distance from web centerline to centroid

Xo Distance from shear center to centroid along the principal Y-axis

MATERIAL TERMINOLOGY SYMBOLS

DW Drywall studs

DWTDL Drywall track: deep leg

DWTSL Drywall track: standard leg

CRC Cold-rolled channel

LG Light gauge

CWS Curtain-wall studs

CWT SL Curtain-wall track: standard leg

CWT DL Curtain-wall track: deep leg

CSJEW C-stud joist extra wide

CSJW C-stud joist wide

CSJ C-stud joist

TBDL Track and bridging: deep leg

NOMENCLATURE

SP Spacing, inches

W Wind load, pounds per square foot (psf)

SIZE DESIGNATIONS

D Section depth, inches

TYPE CWS, HDC, and XCSJ

FY Yield stress = 33 KSI

L/240 Maximum allowable height for a deflection limitation of L/240

L/360 Maximum allowable height for a deflection limitation of L/360

L/600 Maximum allowable height for a deflection limitation of L/600

NATIONAL BUILDING CODES

BOCA Building Officials Conference of America is a nonprofit organization that publishes the National Building Code.

CABO Council of American Building Officials is made up of representatives from three model codes. Issues National Research Board (NRB) research reports.

ICBO International Conference of Building Officials is a nonprofit organization that publishes the Uniform Building Code (UBC).

SBCCI Southern Building Code Congress International is a nonprofit organization that publishes the Standard Building Code (SBC).

UBC Uniform Building Code is the document promulgated by the International Conference of Building Officials (ICBO).

INTRODUCTION

For more than 100 years, cities and skylines have been erected with steel. Time after time, steel's reliability, durability, and consistency have been proven. Today, however, steel has progressed past the world of commercial construction and is now readily accepted by homebuilders across the nation.

The use of steel in residential construction is very much on the rise. Knowledgeable homebuilders and homeowners alike are choosing steel framing for its endless advantages. Steel-framed homes are:

- cost effective
- dimensionally stable
- noncombustible
- termite resistant
- durable
- strong and lightweight
- 100-percent recyclable

In short, steel outperforms wood from virtually every angle. Builders benefit from its consistency, lighter weight, and ease of construction. Homeowners enjoy its durability, superior construction, and flexible building options.

For the residential homebuilder, perhaps the single most important consideration when making the switch from wood-framed to steel-framed construction is the skill of the construction crew. Workers who are not familiar with steel-framed construction techniques require some time, practice, and supervision to become comfortable with, and later skilled at, the new techniques.

Workers who clearly understand the purpose of framing components find the transition fairly easy. Although the attachment methods are different, the basic function of the framing remains the same. In fact, most steel assemblies are easier to accomplish than those made of wood. Typically, workers with only a few hours of experience are able to work efficiently with steel components.

This book attempts to help you make the transition easier. After a study of the table of contents, it is not difficult to see how this book provides you with all the information and simple-to-follow instructions, including more than 350 drawings and photos, that describe how to construct most assemblies. Homebuilders and steel-framing contractors, as well as framer mechanics, often find themselves richly rewarded after they finish their first few steel-framed projects.

CHAPTER 1
INTRODUCTION TO RESIDENTIAL STEEL FRAMING

Technological advances in residential construction have allowed the American home buyer to enjoy steady improvements in construction quality without commensurate increases in cost. In today's economic climate, steel framing has become a cost-effective alternative to wood while upgrading the quality of residential housing and small commercial construction. Frame workers who understand the techniques of steel framing are at the leading edge of today's home-building technology and will provide affordable, high-quality housing for years to come.

Steel framing has long been known for its strength, insect resistance, and fire-protection qualities in commercial construction. It also enables homebuilders to design homes with large, open floor plans (Figure 1.1), straighter walls, quieter partitions, and easier electrical and mechanical installations than that provided by wooden framing. Steel framing also is not subject to the pricing fluctuations common with lumber because steel framing is made from an abundant supply of recycled steel.

Although steel framing might conjure up an image of heavy or cumbersome material, it has emerged recently as the framing material of choice for residential construction. Nationwide, builders and homeowners are recognizing the many benefits steel offers the residential market. Today, cold-formed steel is lightweight, easy to handle, cost effective, and a high-quality alternative to traditional residential framing materials. Steel offers the builder a strong, dimensionally stable, easy-to-use framing system. Figure 1.2 compares a wooden stud with its steel counterpart, the residential metal stud.

Many steel housing components originally were designed as a substitute for wooden framing members. They are now manufactured, however, to reflect the superior strength and consistency of steel. The variety of available steel shapes and sizes has expanded beyond those offered in lumber. This versatility enables the builder to save costs and time while delivering a high-quality product.

Environmental and economic concerns have prompted the building industry to research alternative building materials and methods. This, in addition to its construction benefits and excellent recycling capabilities, is making steel framing the popular choice for resi-

FIGURE 1.1 A steel-framed residential home offers large open floor plans.

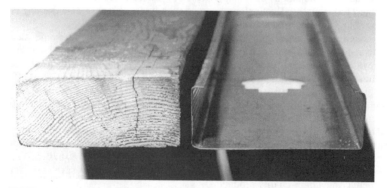

FIGURE 1.2 Comparison of a wooden stud and metal stud.

dential construction. Between 1980 and 1994, steel-framed residential construction grew more than 300 percent. Today, hundreds of builders across the United States and Canada are using steel to frame all types of dwellings: duplexes, resort condominiums, retirement apartments, single family homes, and so on (Figure 1.3).

PUTTING TOGETHER A FRAMING CREW

Perhaps the single most important consideration when contemplating a switch from wood-framed to steel-framed construction is the skill of the framing crew. Builders, constructors, and framers who are not familiar with steel-framing construction techniques might require some time, practice, and supervision to become comfortable with, and later skilled at, the new techniques. Workers who clearly understand the purpose of the framing components find the transition rather easy.

To achieve a proper, quick installation of the steel framework, it is important that the framing team work together. The ideal steel-framing team consists of one knowledgeable steel framer, or lead worker, who can do the layout and cutting and at least two apprentices

A

B

FIGURE 1.3 Steel-framed structures can be (A) two-family duplexes, (B) resort condominiums, (C) retirement apartments, and (D) single-family dwellings.

C

D

FIGURE 1.3 *(Continued)*

who can join together the framing components. Of course, everyone on the team should learn to read blueprints and drawings so that they use the materials where and how they are specified. Workers must be able to distinguish between steel sizes and know which gauges to use in each structural situation. The framer must be able to recognize defects in the steel and know how to make maximum use of the materials on hand without weakening the structure. The team members also must know how to use the tools to cut, trim, measure, mark, and install the framing members.

Perhaps most important of all, steel framers must know the proper methods for assembling and attaching the various pieces that make up the structural framework. An adequate steel framer knows and follows sound steel-framing practices. He or she knows the correct methods to use and has an understanding of why these methods are correct. The lead framer must know not only the methods and the reasons behind them, but also be able to resolve a construction problem without creating problems for others later on the job. The ability to resolve construction problems correctly comes with experience. The experience is useless, however, if the framer does not have the necessary knowledge and understanding of the procedures.

In fact, most steel assemblies are easier than those that use wood. The basic framing function remains the same, although the attachment methods are different. Typically, workers with only a few hours of experience are able to efficiently work with steel. Contractors and builders who recognize that there is a learning curve involved with the transition to steel find themselves richly rewarded after the first steel-framed projects are completed.

DESIGNING FOR STRENGTH AND DURABILITY

Because the strength and durability of any building depends on the steel joists, studs, and rafters in the structural framework, these members must be carefully specified by the designer or architect according to the requirements of the applicable building codes.

It is the responsibility of the designer to specify the size and gauge of the steel for all major steel-framing members. These members must be able to support the loads and stresses that are likely to be put on them during the life of the building. It is the responsibility of the framing contractor to order the framing materials and correctly install them to meet the designer's or architect's specifications.

In wall framing, which includes both exterior walls and the interior partitions between rooms, studs are the main vertical structural members. In roof framing, which includes ceiling joists above the uppermost floor level, the main structural members are rafters. Rafters generally run at an angle to all other framing members. The framing operation is complete when the interior details are framed and the structure is *skinned* with sheathing and ready for the application of exterior and interior surface material.

PROVIDING A COMPETITIVE EDGE

The many advantages to steel framing give the builder or general contractor an edge over the competition. A discussion of some of the more important benefits follows.

Consistent Material Costs. The constant availability of steel means little fluctuation in price. In fact, steel often costs less than lumber. While the price of traditional wooden framing materials has been erratic and grown at a rate much faster than inflation, steel prices have experienced only small quarterly adjustments. The price of steel has been relatively constant over the last decade.

Consistent Quality. Uniform manufacturing tolerances maintain product quality. Steel is always dimensionally correct and does not contain the knots, twists, or warps commonly found in wooden framing materials.

Greater Strength. Steel components with the same outside dimensions as their wooden counterparts are lower in bulk and provide greater strength for many applications (Figure 1.4). The inherent strength of steel also usually translates into a need for fewer members. For instance, less time and labor is needed to space steel studs on 24-inch centers than to space 2-x-4-inch wooden studs on 16-inch centers.

Design Flexibility. A variety of product thicknesses and engineered designs enable steel components to meet specific load requirements economically. Light steel-framed ex-

FIGURE 1.4 Steel offers strength to this hillside home.

terior wall systems can be used on buildings of almost any shape. They readily adapt to flat, curved, or angular surfaces and can be incorporated into recesses or projections (Figure 1.5).

Ease of Installation. Steel studs install quickly. The components of a lightweight, steel-framed assembly, including studs, track, joists, etc., can be, and frequently are, assembled quickly and easily into prefabricated panels or assemblies (Figure 1.6).

Time and Cost Savings. The majority of owners, developers, and construction managers naturally relate construction time to cost. Since construction financing is expensive, its time duration must be minimized. Everyone concerned with financing a project also is aware that the return on the investment does not begin until the building is completed and occupied.

Ease of Modification. Out-of-place studs can be moved easily to assure the accurate attachment of wallboard and other construction components.

Ease of Utility Installations. Spaced at regular intervals, the punch-outs in steel-framing members serve as raceways for electrical and plumbing lines (Figure 1.7). Prepunching allows plumbers and electricians to rough-in their work quicker than they can drill holes through wooden framing members. This benefit reduces job costs. Grommets or conduits protect wiring from the sharp steel edges of the punch-outs.

Custom Precuts. Many steel members are available in a variety of precut, standard shapes and sizes, as well as custom shapes and sizes. This feature minimizes construction waste.

Superior Sound Control. The resilient nature of the steel studs helps walls absorb sound. This is especially true when combined with sound attenuating insulation.

Additional Fire Safety. Steel studs are noncombustible and improve fire safety compliance with local codes and fire regulations. In fact, history shows that construction with light steel framing can lower insurance premiums.

Improved Construction Quality. Galvanized steel studs are corrosion resistant and dimensionally stable, which eliminates nail popping and reduces cracking, both of which of-

FIGURE 1.5 Steel-framed curved projection.

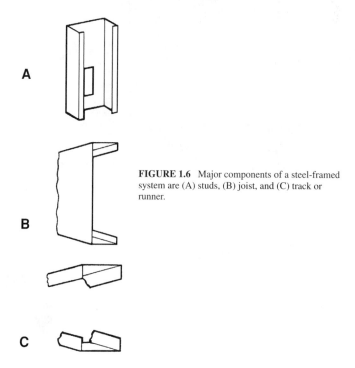

FIGURE 1.6 Major components of a steel-framed system are (A) studs, (B) joist, and (C) track or runner.

ten are associated with the shrinkage and expansion of wooden materials. Steel studs do not warp, split, rot, or attract termites.

Labor Savings. Steel-stud systems generally are simpler to erect and errors are easier to correct than with wooden systems. Once crews are experienced with steel framing, labor

FIGURE 1.7   The holes in the studs serve as a raceway for the installation of electrical and plumbing lines.

costs and time are reduced significantly when compared to the construction of wooden systems.

User-Friendly Technology. Steel framing is a proven, user-friendly technology that allows for a smoother transition from other materials.

Reduced Weight. Steel members weigh as much as 60 percent less than wooden members, which allows foundation and even seismic loads to be reduced (Figure 1.8).

Multiple Finishes. Steel framing accommodates all types of commonly used finishing materials. With lightweight steel framing as the exterior wall backup, the variety of finishing materials, textures, and shapes that can be selected is virtually unlimited. The exterior-insulation-finishing system (EIFS) works well with steel framing.

Availability. Steel framing components are available throughout the country. They can be purchased in stock lengths, custom-cut dimensions, or in preengineered, panelized systems. Because of the growing use of steel in residential construction, local lumberyards and commercial building-supply warehouses are adding steel-framing components

FIGURE 1.8 Steel framing, such as this roof truss, are lightweight when compared with wooden trusses.

to their inventory. Sometimes builders also can purchase steel components directly from the manufacturer.

Building Code Compliance. While building codes usually restrict wooden framing to three stories, but the noncombustibility and high strength of steel combine to allow load-bearing construction to six stories (Figure 1.9). As a result, land-use densities can be increased to produce greater returns on investment for developers. For builders of multiple

FIGURE 1.9 Most building codes allow steel construction up to six stories.

units, this factor becomes increasingly important as land prices and construction costs continue to rise.

With its excellent strength-to-weight ratio, light steel framing increases development opportunities by permitting load-bearing, midrise construction on soils that have poor bearing capacities, without costly oversized footings. This is a distinct advantage over post-and-beam and heavy concrete-block construction methods. This attribute also produces an economical backup system for curtain-wall applications with resulting dollar savings in spandrel beams, columns, and footing sizes.

Light steel framing affords the builder the benefits of faster installation, faster enclosure, faster occupancy, reduced interim financing, and a quicker return on investment.

BENEFITING THE HOMEOWNER

In order for a builder or contractor to take advantage of the benefits of steel framing, he or she must be able to sell steel framing to prospective homeowners. To do this, it might be helpful to point out the many benefits homeowners can realize from the purchase of a steel-framed house.

Inherent Strength and Noncombustibility. Steel's inherent strength and noncombustibility enable a steel-framed house to resist such devastating events as fires, earthquakes, and hurricanes. Homes can be designed to meet the greatest seismic and wind-load specifications in any part of the country. In fact, unlike wood-framed residences, homes constructed with steel frames have a distinct advantage when it comes to earthquakes. Properly built homes that utilize steel framing can be engineered to receive a seismic 4 rating, which is the highest rating for residential dwellings.

The reason steel framing can meet these high standards is not only steel's strength and resiliency, but also its quality of assembly, positive attachment methods, consistent quality without defects, and the fact that it is securely bolted to the foundation. As insurance investigations show, most earthquake damage to wood-framed homes results from the house separating from the foundation, which is something that usually does not occur with steel-framed residences.

NOTE: When Andrew, the worst hurricane in U.S. history, ripped its way across southern Florida in October 1992, buildings toppled like cardboard boxes. Metal roofing and steel-framed housing, however, withstood the fury better than most conventional products. Some structures built with metal products were left virtually unscathed (Figure 1.10).

Common Appearance. Once finished, homeowners cannot tell the difference between a wood- and steel-framed house. Once the steel frame is up, it can be finished in wood, stucco, brick, vinyl siding, or any other finishing material (Figure 1.11A, B, C & D).

Resistance to Termites and Pests. Steel is impervious to termites and other wood-destroying insects. Steel-framed homes provide health benefits to the homeowner and the community because they do not require pesticides or the application of other chemicals used to treat wooden framing products. In fact, steel framing is recommended by the Healthy House Institute and several similar organizations for chemically sensitive and environmentally conscious homeowners who seek good indoor air quality. Steel frames do not need to be treated for termites and are free of resin adhesives and the chemicals normally present in other construction materials.

Ease of Remodeling. Since steel framing allows for larger spans, a home can be designed without interior load-bearing partitions. This benefit makes it easier for homeowners to complete alterations without affecting the structure. Walls that don't bear a load can be easily removed, altered, refit, and relocated (Figure 1.12). Outbuildings, garages, and sheds can be framed as independent structures (Figure 1.13).

FIGURE 1.10 Hurricane damage to steel-framed house was only to a small portion of the roof.

Rust Protection. Galvanized steel framing components protect the home against rust. Even steel-framed homes built close to the seashore do not have problems with rust.

Improved Resale Value. Since a steel-framed house does not look any different than the neighboring wood-framed house, the steel-framed house should sell just as easily. Because of steel's high strength and durability, the home should last and retain its value for a long time. In fact, if the builder takes advantage of steel's strength and flexibility by designing wide open spaces, the house should have additional selling features.

Flexible Designs. The strength of steel enables homes to be designed with larger open spaces. This flexibility gives the architect the option of building any kind of home from

A

FIGURE 1.11 Steel-framed residential homes can be finished in almost any exterior finishing material including (A) wood, (B) stucco, (C) brick, and (D) vinyl siding.

B

C

D

FIGURE 1.11 *(Continued)*

FIGURE 1.12 Adding an extra room is no problem when using steel framing.

FIGURE 1.13 Outbuildings like this large storage shed can quickly be erected using steel-framing techniques.

traditional to contemporary to avant garde. Steel-framed homes can be finished with virtually any kind of material and can easily be remodeled or enlarged.

Noncombustible Materials. As mentioned earlier in the chapter, steel is noncombustible and provides improved fire safety compliance with local codes and fire regulations. In fact, construction with lightweight steel framing actually might lower insurance premiums.

Energy Efficiency. Energy conservation is a concern that all of us share. When comparing steel-framed exterior walls with other systems, there is little doubt about which system best reduces energy loss. The energy transmission through a wall is directly related to the resistance (R) provided by the construction. Increasing the R value, or resistance, means less heat is transmitted. Whether trying to prevent interior heat loss in a cold climate or interior heat gain for an air-conditioned building in a warm climate, the need to increase the R value is the same. When insulated, as described in Chapter 9, steel structures provide excellent energy efficiency (Figure 1.14). An additional benefit is the fact that because steel is dimensionally stable, it does not expand or contract with humidity changes.

DISAVOWING THE MYTHS

A steel frame does not cause interference with television or radio reception, or with garage door openers. There is no problem should lightning strike the home because the steel framing provides multiple conductive paths directly to the ground. There is no additional noise, and there is no problem putting pictures on walls and installing hanging lamps or plants because you simply use metal screws instead of nails. That should dispel the myths.

FIGURE 1.14 Installing insulation in between metal studs.

PROTECTING THE ENVIRONMENT

Experts in many fields are looking for new building strategies and technologies that will help sustain affordable growth into the next century. Steel is one of those building materials. By using recyclable materials in the structures we build today, we can recycle tomorrow. Our descendants will be able to reuse the steel we use today.

Steel producers are steel recyclers. Steel can be recycled an infinite number of times without degradation. During the last decade, more than 500 million tons of steel scrap were recycled. More steel was recycled during this period than any other material, which kept this valuable commodity out of the nation's dwindling landfills. Magnetic separation makes steel the easiest and most economical material to remove from the solid-waste stream. It ensures reclamation of recyclable steel for delivery to scrap dealers, detinners, and mills.

The steel industry consumes twice the amount of recycled material as all other industries combined. By using recyclable materials, such as steel, for today's homes, we can ensure the availability of raw materials, such as wood, for other future applications. The use of steel in residential construction can help us achieve the American dream of affordable housing well into the future.

The North American steel industry manages approximately 400,000 acres of land to satisfy its mining needs. This is 1 percent of the land needed by the timber industry to meet the current demand for its wood-based products. The iron ore mines in Minnesota, alone, can satisfy North America's steel needs for hundreds of years. Steel producers are becoming more efficient at producing more finished goods using less raw materials.

Scientists and engineers are finding new ways to design and use steel more effectively and efficiently. Today's automobile, for example, requires only 50 percent of the steel used in models produced in the 1960s. Steel cans are 30 percent lighter and steel mills improve their yields annually. Tomorrow's house will be designed with an efficient steel frame that uses 20 to 30 percent less steel than today. Steel continues to reinvent itself.

Regardless of the issue—economics or the environment, educated consumers or efficient construction methods—the benefits associated with steel are gaining acceptance for residential steel-framed systems. When it comes to conserving the earth's resources, you cannot beat the one tremendous advantage steel has over wood: it saves forests. After all, when you build a wood-framed house you kill trees. On the other hand, when you build a steel-framed house you recycle old Fords and Toyotas.

ASSEMBLING STEEL-FRAMED RESIDENCES

There are three basic assembly methods for steel-framed residential construction:

- stick-built construction
- panelized systems
- preengineered systems

Building Piece by Piece

Stick-built construction is virtually the same for wood and steel. This framing method, pictured in Figure 1.15, has actually gone through a transformation that incorporates many of the techniques used in panelized construction. The steel materials are delivered to the job

FIGURE 1.15 Installing load-bearing steel framing with the stick-built construction system.

site in stock lengths or, in some cases, cut to length. The layout and assembly of the steel framing is the same as that for lumber, except the components are screwed together rather than nailed. Steel joists can be ordered in lengths long enough to span the full width of a home. This expedites the framing process and eliminates lap joints. Sheathing and finishing materials are fastened with screws or pneumatic pins.

Stick-built construction is an economic alternative. It is most efficient when weather is not a factor, when the transportation of prefabricated panels from the contractor's shop is burdensome, or when there is not enough repetition in the size and shape of the panels to realize a cost savings from prepanelization. Stick-built construction easily conforms to tolerance problems or changes at the job site.

Using Panelized Systems

Panelization is a system that prefabricates walls, floors, and/or roof components into sections (Figure 1.16). This construction method is most efficient when there is a repetition of panel types and dimensions (see Chapter 4). Panels can be made in the shop or in the field. A jig is developed for each type of panel. Steel studs and joists are ordered in cut-to-length sizes for most panel work. They are placed into the jig and then fastened using screws or

FIGURE 1.16 Installing a panelized wall.

welding (Figure 1.17). The exterior sheathing, or in some cases the complete exterior finish, can be applied to the panel before it is erected.

Shop panelization can offer the builder several significant advantages. The panel shop provides a controlled environment where work can proceed regardless of weather conditions. The application of sheathing and finishing systems is easier and faster with the panels in a horizontal position. Although the panels must be transported from the panel shop to the job site, most often the cost advantages offset the added transportation costs.

A major benefit to panelized systems is the speed with which they can be erected. A job usually can be framed in about one-fourth the time required to stick-build a structure. When the exterior finishing system also can be part of the panel, the overall time savings can be even greater.

Another important consideration for many builders is the fact that completed panels can be erected in below freezing temperatures. This permits year-round construction in all areas of the country. Since there are minimal delays, construction normally can be completed on schedule. Advance fabrication can be accomplished in a temporary shelter at the site, or in the contractor's shop ahead of the construction schedule, and then stockpiled and delivered to the site. There are no shrinkage, warping, or swelling problems with prefabricated steel framing as there might be with wood that has been exposed to the elements at the job site. The panel size that was designed and constructed is the panel size that is installed. As a result, material damages due to adverse weather conditions are negligible.

Quality control is better, too. The panels are assembled in one place by an experienced crew working with the right equipment under controlled conditions (Figure 1.18). Panels can be inspected before they are installed. If there are any problems, the entire job schedule is not disrupted.

FIGURE 1.17 Connecting the frame studs by welding.

FIGURE 1.18 Up goes a load-bearing wall with a little help from the work crew.

Attaching Preengineered Systems

Steel's great strength and design flexibility makes innovative systems possible. Preengineered systems typically space the primary load-carrying members on more than 24-inch centers, sometimes on as much as 8-foot centers. These systems use either secondary horizontal members to distribute wind loads to the columns or lighter weight steel in-fill studs between the columns. Furring channels that support sheathing materials also provide a break in the heat-flow path to the exterior, which increases thermal efficiency.

Many of the preengineered systems provide framing members that are precut to length and have predrilled holes for bolts or screws (Figure 1.19). Most of the fabrication labor is done by the supplier, which allows a home to be framed in as little as one day.

FIGURE 1.19 The framework of a preengineered steel-framed home (top); the completed preengineered home (bottom).

SELECTING FRAMING COMPONENTS

The anatomy of a steel-framed house is basically the same as one built with wood (Figure 1.20). Steel framing picks up the load and then transfers and distributes it just like wooden framing. Both steel- and wood-framed systems incorporate joists, beams, studs, headers, trimmers, cripples, etc.

As discussed in this book, there are some important, basic differences between steel- and wood-framed systems that need to be addressed before the builder or contractor can decide which system is appropriate for his or her particular situation. While steel-framing materials are generally less expensive than wood and require less time to assemble by a knowledgeable work force, other costs might be incurred such as on-the-job training and engineering fees for truss systems, load-bearing walls, and certain joist assemblies (see Chapter 6).

Light steel-framing components are engineered products that meet a variety of load requirements. The engineered nature of the product enables structures to be designed to meet the exact load needs. The amount of engineering support required depends largely on where the framing is used.

It is important to recognize, however, that steel-framing components are designed to be compatible with wood. As a result, many builders find it convenient to use a combination of steel and wood until their workers feel comfortable using steel exclusively.

There is a cold-formed steel-framing member for virtually any situation. This simplifies the framing design and provides maximum structural efficiency, design flexibility, and ease of installation (Figure 1.21). The construction elements of a steel-framed house are divided into three main assemblies: floors, walls, and roofs. While these assemblies often are consistent with wood-framed construction, some steel assemblies also might involve new and innovative framing systems.

The steel component known as the structural C is the predominant shape of studs and beams (Figure 1.22). Steel's greater strength offers wider spacing between members and longer spans. This feature increases design flexibility and decreases material and labor costs. Steel members commonly are spaced on 24-inch centers. With steel's great strength, however, some innovative designs space the members up to 8 feet on center using horizontal furring channels at on 24-inch centers to tie the system together (Figure 1.23).

BUILDING IN PHASES

There are four framing phases that can use steel components: floor joists, exterior and load-bearing walls, nonload-bearing partition walls, and roof rafters and supports.

FIGURE 1.20 Basic anatomy of a steel-framed house.

FIGURE 1.21   The steel structure framed and ready for finishing (top); the finished house (bottom).

Supporting Floors

Generally, builders opt for joist depths that range from 6 to 12 inches with a steel thickness from 0.034 to 0.101 inch. Instead of using lapped joists on multiple span conditions, a single, continuous length of joist commonly is used. Steel joists offer distinct advantages for residential construction because they come in sizes and gauges that exactly meet antici-

FIGURE 1.22 The C structural wall studs are bundled when sent to a prefab shop or the job site.

FIGURE 1.23 The 48-inch on-center system is versatile, reduces costs, and makes assembly easier.

pated load and deflection requirements. They also can span longer distances, which allows architects and builders to design open floor plans (Figure 1.24).

While steel joists provide a great deal of overall strength, that strength can be diminished greatly if the web is not reinforced at points of stress, such as the joist ends that rest on the foundation or the I beam. Certain structural features, such as balconies or overhangs, require similar reinforcement. The web of the joist typically can be reinforced at those points with either web stiffeners or sections of steel runner screw-attached in place. Joists must meet the load requirements spelled out in local building codes and must take into account the added load from tile and other floor coverings.

Most steel frames in residential construction are laid out using both 16- and 24-inch centers, although 24-inch centers are optimum. The ends of the joists fit into track material that takes the place of wooden rim joists. The joists are fastened to the track by driving screws through the flanges and installing a screwed clip that attaches the joist web to the track. The area then can be reinforced by installing web stiffeners, which are vertical pieces of stud material that are screwed to the web of the joist.

Solid blocking and flat steel strapping reinforce joists midspan between the carrying beams and partitions. If the span is more than 14 feet long, the American Iron and Steel Institute (AISI) recommends more rows of reinforcement materials between the supports. Install the blocking between the joists, pull the flat strapping tightly across the tops and bottoms of the joist, and then fasten the blocking to the joists. Solid blocking also can be used to reinforce floor joists that carry walls and other loads.

Glue and screw-attach plywood subflooring to the steel framing. A couple of pneumatic nailers can fasten plywood to steel, but the nailgun manufacturers recommend initially setting the plywood with screws and then using their products to back-nail the sheets.

Building Nonload- and Load-Bearing Walls

Use structural studs for both interior and exterior load-bearing walls (Figure 1.25). The structural studs that are used in wall construction range in size from 2½ to 8 inches and in metal thickness from 0.034 to 0.071 inch. The metal studs are available in sizes from 1⅝ to 6 inches and have a metal thickness from 0.018 to 0.034 inch. Although 3⅝- and 6-inch walls are common in stick-built residential construction, the wall thickness can be varied

FIGURE 1.24 The flooring shows the size of one room that can be framed around the parameter.

FIGURE 1.25 The exterior walls of a structure that hold its weight are load-bearing walls (top), while the interior framing walls in most cases are the nonbearing walls (bottom).

to account for insulation requirements. Exterior insulation board also can be applied to the walls to increase thermal efficiency instead of increasing the cavity dimension.

As a rule of thumb, 2½-×-4-gauge, 3⅝- or 3½-inch steel studs are used, except where cabinets are located. The only required change is to use screws instead of nails. Mount the top and bottom runners where partition walls are located. Twist the studs in place and then crimp or screw them into position.

Door openings, for example, often are framed with wood to provide a nailable backing for jambs and casings. Wooden blocking also can be screwed into position for jambs and casings. Wooden blocking also can be screwed to the framing between studs to help mount cabinetry or crown molding. A wooden baseplate or wooden blocking can be inserted into the bottom runner as a nailer if the baseboards are not attached with adhesive or screws.

The uniformity of steel framing makes it superior to wooden framing for load-bearing construction. Steel framing has no knots, cracks, warps, or twists, which ensures reliable performance. With steel's wide range of size and gauge options, studs can be matched to meet even the most difficult requirements for both wind and axial loads. The wall-framing quality also benefits from straight steel studs. Framers use layouts of both 16- and 24-inch centers, with the optimum spacing being 24 inches. Steel studs generally come in standard lengths from 8 to 12 feet, although they can be obtained in almost any length (see Chapter 2).

The studs fit into light-gauge steel tracks that serve as top and bottom plates. The framer screws through the track flanges and into the studs to fasten them together. Unlike wooden plates, however, the tracks are not structural and do not help span floor irregularities. That is why steel framers say it is more important to get steel floors flat and level than it is wood-framed floors.

Depending on the type of house being built, steel framers either stick-build the walls and then attach the exterior sheathing (Figure 1.26), or they assemble the walls and attach the sheathing before tilting up the wall sections.

The type of exterior sheathing and the insulation used is important to the energy performance of steel-framed houses. In some climates, industry experts recommend rigid insulating sheathing. For complete details on the thermal transfer—energy saving—of steel-framed houses, see Chapter 8.

The details involved with bracing and shear walls are among the most important aspects of these wall systems. They commonly call for flat steel strapping to be run diagonally

FIGURE 1.26   Wall sheathing adds strength to the structure.

across interior and exterior walls and then fastened to the studs. Secure the strapping to the foundation or slab and then run it, under tension, all the way to the top track. You also can use plywood or steel sheets screwed to studs to create shear walls. Because most wall tracks are not structural, it is important to align the framing. It is especially important to position the rafters over the studs.

Designing Roofs

Steel framing can be used in virtually any roof system, from the simplest shed roof to the most complex hip-and-valley roof system. The conventional ridge-board/rafter system (Figure 1.27) or trusses can be built onsite or offsite in truss plants, in completed or sectioned assemblies (Figure 1.28).

Building roof-truss assemblies to meet specific loads requires special planning and skill. Assembly might require miter cuts, gusset plates, special attachments, and, in some cases, welding. This is the last area of construction to be considered in the transition between wood and steel.

The roof finishing material used on the steel-framed roof depends on the owner's or contractor's selection. It can be metal, asphalt or wooden shingles, slate tile, thermoplastic, etc. (Figure 1.29).

CALCULATING THE COSTS

The real value of switching from wooden to steel framing often is difficult to calculate, especially in the early stages when workers might be unfamiliar with the use of the steel materials. That is why many experts in the field recommend that builders first use steel for

FIGURE 1.27 Conventional ridge-board/rafter system.

FIGURE 1.28 Typical truss construction.

nonload-bearing interior partitions to minimize out-of-pocket costs and construction risks (see Chapter 6). This also can be a way to instill worker confidence and familiarity with light steel framing.

Table 1.1 provides steel framers with a way to determine the out-of-pocket cost differentials between wood-framed and steel-framed building methods.

FIGURE 1.29 Various roofing materials can be used in steel roof framing. They also can be used for sidewalls.

FIGURE 1.29 *(Continued)*

TABLE 1.1 Steel Framing Value Analysis Calculator

Nonload-bearing interior partition walls:

Calculate the lumber requirements for interior nonload-bearing walls in your normal fashion and enter the totals here.

Studs	_____
× cost per stud	$_____
Total cost of wood	$_____

Calculate the light steel framing requirement as follows:

Studs (vertical only)	_____
× cost per stud	$_____
Total stud cost	$_____

Calculate the runner requirements as follows (linear feet of walls ÷ 5 = total top and bottom runners):

10" standard length	_____
× cost per runner	$_____
Total runner cost	$_____

Calculate total steel needs.

Studs	$_____
+ runners	$_____
Total steel costs	$_____

Calculate the cost differential between steel and wood.

Steel costs	$_____
Lumber costs	$_____
Differential	$_____

Adjustment for first time equipment purchases:

Chopsaw	$_____
Abrasive metal-cutting blade	$_____
Crimpers	$_____

TABLE 1.1   Steel Framing Value Analysis Calculator (Continued)

Nonload-bearing interior partition walls:

Screwgun	$_____
Subtotal	$_____
Total first-time cost to switch	$_____

Calculate cost differential of floor joists.

Wooden floor joists	_____
× cost per joist	$_____
Total cost of wooden joists	$_____
Steel floor joists	_____
× cost per joist	$_____
Total cost of steel joists	$_____

Calculate costs of steel runners.

Linear feet of foundation perpendicular to joists, ÷ 10	_____
× cost per runner	$_____
Total cost of steel joists	$_____
Cost differential between steel and wooden joists	$_____

Load-bearing walls
Calculate the lumber requirement for interior load-bearing walls in your normal fashion and enter the totals here.

Studs	_____
× cost per stud	$_____
Total cost of wood	$_____

Calculate your light steel framing requirement as follows:

Studs (vertical only)	_____
× cost per stud	$_____
Total cost of steel studs	$_____

Calculate runner requirements.

Linear feet of walls ÷ 5 = total top and bottom runners	_____
× cost per runner	_____
Total runner cost	$_____
Total steel cost	$_____
Cost differential between steel and wood	$_____

CHAPTER 2
MATERIALS AND TOOLS

The design of any steel residential structure must take into account the possible load conditions and the resulting stresses and movements. Load-bearing walls must be designed to carry the weight of the structure, its components, and the other loads that occur once the building is occupied. The amount of axial load that structural members can bear varies with the amount of lateral load, which is the pressure from wind or other horizontal forces that the final assembly might incur.

Manufacturers of residential steel-framing components provide tables and charts in their catalogs (Figure 2.1) that identify such items as the physical and structural properties of the steel-framing materials, wind-loads, floor and roof applications, web crippling, strength figures, limiting-heights, and so on. When ordering steel-framing components, keep in mind the following.

* Interior nonbearing partitions are not designed to carry axial loads.
* Limiting heights are based on stress or deflection limits for a given lateral load.
* Height limitations depend on the gauge of steel used, the dimensions of the stud, stud spacing, and the allowable deflection limit.
* The load-span capacity of steel studs is based on the following factors when applicable:
  ~American Iron and Steel Institute (AISI) specifications for the design of cold-formed structural steel members
  ~yield strength
  ~sectional, structural, and physical properties of members
  ~bending stress
  ~axial stress
  ~shear stress
  ~allowable deflection
  ~web crippling at supports
  ~lateral bracing ratio
  ~wind speed or wind load

FIGURE 2.1 Catalogs from steel manufacturers.

NOTE: Sample tables and charts from the catalogs of well-known steel-framing manu-facturers are included in Appendix B. The List of Acronyms includes common symbols from the catalogs and their meanings.

The performance of any residential steel-framing system must comply with the regula-tory requirements established by local, state, and federal agencies. Be sure to consider lo-cal and state building codes, as well as insurance and lending agency requirements before selecting residential steel-framing systems. The selected appropriate framing should re-flect the system limitations.

Such structural factors as limiting height and span, required number of screws, metal thickness, bracing spacing, or maximum frame and fastener spacing should not be ex-ceeded since they affect the flexural properties and strength of an assembly. The yield strength of all steel is not the same. Substitution by size alone is not recommended.

NOTE: The system performance after any material substitution or compromise in as-sembly design cannot be certified and might result in failure under critical conditions.

FOLLOWING GENERAL FRAMING REQUIREMENTS

Selecting the appropriate framing materials depends on a number of factors. In the case of wooden framing, these include the species, size, and grade of the lumber. In the case of steel-framing members, configuration, size, thickness, and grade of steel must be considered. Equally important are frame spacing and the maximum span of the surfacing material.

Select framing members and the appropriate installation method according to their ability to withstand the loads to which they might be subjected. These include live loads, which are contributed by occupancy and elements such as wind, snow, and earthquake, and dead load, which is the weight of the structure.

Meeting Deflection Criteria

Even though an assembly is capable structurally of withstanding a given load, its use might be restricted because the amount of deflection that occurs when the load is applied exceeds that which the surfacing materials can sustain without damage. Obviously, this de-flection factor influences the selection of surfacing materials.

Deflection is defined as the difference in stud displacement at midspan when the stud is under pressure compared to when there is no load. The deflection limit is the total height of the stud in inches divided by the deflection criteria. For example, if the specification requires a $\frac{1}{360}$ on a 10-foot-high wall, convert the 10 feet into inches and divide by 360. The result is $\frac{120}{360}$ or 0.33 inch. This means the wall cannot deflect more than $\frac{1}{3}$ of an inch at midspan.

There are different deflection criteria for different jobs. The higher the deflection criteria, the stiffer the wall must be. There is a big controversy, however, about what is the appropriate deflection criteria for different building heights and veneers.

For a one- or two-story building with brick veneer, stucco, or other brittle surfaces, L/360 is adequate. For a multistory brick building, L/600 must be used. For exterior insulation finishing systems (EIFS) or wooden and metal siding, L/240 is adequate.

Allowing for Section Properties

Moment of inertia and section modules are functions of the shape and thickness of the material. In general, the bigger the stud webs, flanges, and return lips, the higher the values must be. The same holds true for gauges. When the specifications request the minimum on these properties, refer to the manufacturer's literature to find a stud that meets or exceeds the minimum.

When specifications call for a C-shaped load-bearing stud or joist, it might not be enough to let the contractor know what type of stud to use. Manufacturers have come up with different C-shaped studs that have different flanges and returns. If the specifications do not specify the exact flange size, any of the various C-shaped studs can be used as long as they meet the design loads and criteria.

Specifying Yield Strength and Grades of Steel

Yield strength is usually taken to be that stress that leaves the steel with a permanent elongation from 0 to 2 percent after the stress is relieved. A stronger steel has a higher yield. Specifications might call for the steel to have a yield strength of 33, 40, or 50 KIPs per square inch (ksi), with a KIP equal to 1000 pounds. To determine if the requirement is higher than the manufacturer's standard yield strength, refer to the manufacturer's literature. There might be an additional charge for higher yield material.

The specifications also might call for certain American Society for Testing and Materials (ASTM) requirements for different grades of steel. They also might call for a certain type of coding. The steel grade refers to the yield strength: grade A is 33 ksi, grade C is 40 ksi, and grade D is 50 ksi. The coding, expressed by a G number, refers to the amount of zinc in the steel. Higher values mean that there is more zinc per square foot. Many manufacturers provide G 40 and G 60 as standard and charge more for G 90 steel. Before asking for a bid from a supplier, obtain specific information so that you get a realistic dollar cost. The type of stud, return, flanges, and coding, ASTM requirements, and yield strength can have a sizable impact on price and must be known in advance and provided to the manufacturer when a quote is requested. Some manufacturers have in-house technical personnel that might be able to assist in the selection of studs, but they are not likely to go into great detail.

Figuring in Wind Speed and Wind Load

Wind speed is a primary factor in figuring wind load. Almost every building code has a map that depicts the average of the highest wind loads in each area during the past 50 to

100 years, per the American Society of Civil Engineering's (ASCE) code. Architects are required to design a building for worst-case situations, per the relevant building codes. All codes have a different way of doing this, but there is usually a table for converting miles per hour (mph) to pounds per square foot (psf). There are also several other factors to consider, such as pressure coefficient(s) and gust factor(s).

Taking into consideration all the factors is not a simple conversion of one to the other. In order to make some of the decisions, you must be an engineer. For instance, let's look at how you might choose an exposure factor for a particular building based on the location of the project and the requirements of just the Uniform Building Code (UBC), which is one of the many factors that must be considered.

If the building is set on top of a hill, has no other buildings around it, and has no trees within one mile, the appropriate exposure factor is C, which is the most critical.

In most cases, the design engineer selects the worst-case situation because he or she does not know the terrain surrounding the project. This is where a contractor's knowledge can be helpful to the engineer. The contractor can help the engineer select an exposure that is closer to reality. This in turn might help the contractor financially, since a realistic exposure might allow him or her to use a lighter gauge steel. This is where past experience is helpful. If the contractor only bids one- or two-story shopping strips in a particular city or area, he or she could assume that the wind pressures are the same as in previous jobs.

Spacing Frames

Frame spacing is a factor in determining load-carrying capabilities and deflection. It also can be a limiting factor for the finishing materials. Every finishing or surfacing material is subject to a span limitation. This is the maximum distance that a material can span between framing members without undue sagging. Some manufacturers include maximum frame spacing tables for the various board products in their catalogs. When frame spacing exceeds the maximum limits, install furring members to provide necessary support for the surfacing material.

Installing Insulation and Utility Services

Plumbing, electrical, and other fixtures and mechanicals within the framing cavities must be flush with or inside the plane of the framing (Figure 2.2). Drive the fasteners reasonably flush with the surfaces. In wood-framed construction, attach the flanges of batt-type insulation to the sides of the framing members and *not* to their faces. Any obstruction on the face of the framing members that prevents firm contact between the finishing material and the framing can result in a loose or damaged surface and fastener imperfections.

Withstanding Bending Stress

Steel-framing members also must withstand any exerted unit force that breaks or distorts the frame stud, based on the capacity of the stud acting alone. This action is called *bending stress*.

FIGURE 2.2 Utility services should fit within the framing and must be flush with framing surfaces.

Planning for Structural Movement

While meeting most current building-design standards, residential steel frames generally are more flexible and offer less resistance to structural movement. This flexibility and the resulting structural movement can produce stresses within the usually nonload-bearing wall assemblies. When these accumulated stresses exceed the strength of the materials in the assembly, they seek relief by cracking, buckling, or crushing the finished surface, unless relief joints are provided to isolate these building movements.

Structural movement and most cracking problems are caused by deflection under load, physical changes in the materials due to temperature and humidity changes, seismic forces, or a combination of these factors.

Allowing for Concrete Floor Slab Deflection

Dead and live loads cause deflection in the floor slab. If this deflection is excessive, cracks can occur in the partitions at the midpoint between the supports. If possible, wait about two months after the slabs are completed to install the partitions. This allows up to two-thirds

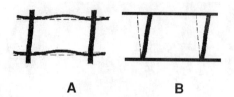

A   **B**

FIGURE 2.3 Structural movement: (A) floor-slab deflection (noncyclical) and (B) racking (cyclical).

of the ultimate creep deflection to take place and reduces the chances of partition cracking. Creep deflection is usually a onetime, noncyclical movement (Figure 2.3A).

Figuring in Wind Load and Seismic Forces

Wind and seismic forces cause shearing on the building framework, which distorts the rectangular shape to an angled parallelogram. This distortion, called *racking*, can crack and/or crush the partitions adjacent to columns, floors, and structural ceilings. This is a cyclical movement that occurs with the given force (Figure 2.3B).

Wind blowing against a building exerts a positive pressure force on one side of the building and a negative pressure on the opposite side. Because the building is secured to the ground, the greatest racking occurs at the upper floors.

In an earthquake, the ground rocks, twists, heaves, and subsides, all the while changing direction and speed. Such violent and chaotic ground movement sets buildings in motion. Houses tend to shift off their foundations. Houses literally come apart at the seams, section by section and piece by piece. Some structures even overturn (Figure 2.4). Steel-framed houses, however, if properly attached to the foundation and tied together structurally, can resist seismic forces and reduce the likelihood of earthquake damage.

The light weight of steel-framed buildings results in less force from inertia. Less force means less damage (Figure 2.5). Steel's natural flexibility also is an advantage during seismic events. The screwed or welded joints in steel-framed buildings also help dissipate energy and motion. To be effective, however, steel's inherent earthquake resistance must be accompanied by design and construction techniques that take advantage of these characteristics.

Prefabricated structural panels that are screwed or welded to wall framing add rigid bracing and help resist lateral loads and keep the house in one spot. The racking effect also can be reduced by stiffening the frame with shear walls and/or crossbracing. Diagonally brace light steel-framed buildings with steel strapping. Larger buildings resist racking with shear walls and bracing, in addition to the strength added by the finishing materials.

Isolate partitions from the structure to keep them from cracking as a result of the racking movement and distortion. Securely connected wall, floor, and roof frames also help tie a house together and make it a single, solid structural unit. Proper connections do more to hold a house together during an earthquake than any other single seismic design element.

Modern building codes require seismic design elements in new construction. Those elements typically include the measures mentioned above. Consult your local building codes for the requirements in your area. Older houses frequently need retrofitting if they are to withstand earthquakes. While this book deals primarily with new construction, the same principles apply to retrofit applications.

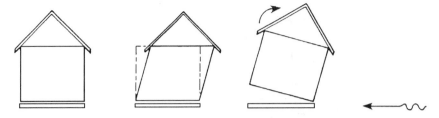

FIGURE 2.4 The effect of a lateral force on structural elements.

SELECTING STRUCTURAL COMPONENTS

Steel framing components are designed to be compatible with wooden framing materials. The standard dimensions of a 2-×-4-inch wooden stud, for example, are now 1½ × 3½ inches (formerly 1⅝ × 3⅝ inches). In use, there can be ⅛ inch or more variation in wooden studs for saw-cut differences, warping, or twisting.

The main structural components of the lightweight steel-framing system—the stud runners—usually are channel-type, roll-formed from corrosion-resistant coated steel, and designed so that the facing materials can be screw-attached quickly. They are strong, nonload-bearing components of interior partitions, ceiling and column fire-proofing elements, and framing for exterior curtain-wall systems. Members with a heavier thickness are used in load-bearing construction. Limited chaseways for electrical and plumbing services are provided by punch-outs in the stud web. Match the runners for each stud size. Then align and secure the studs to floors and ceilings, where they also can function as headers. The factory often color-codes studs and runners on the end to indicate gauge and help identify products at the job site.

When selecting a stud gauge (Table 2.1) and size, you must take into account a number of factors. The key consideration is whether the assembly is for a load-bearing, nonload-bearing, or curtain-wall application. Other variables include anticipated wall height, the weight and dimensions of mounted fixtures, the desired fire rating, sound attenuation needs, anticipated wind loads, insulation requirements, deflection allowance, and the desired impact resistance.

In general, stronger or heavier studs are needed to accommodate taller walls. Stronger studs also reduce deflection and vibration from impacts such as slamming doors. Wider studs might be required to handle insulation requirements. Fire-rated buildings typically are designed, tested, and approved based on the lightest gauge, shallowest stud depth, and maximum stud spacing indicated in the assembly description. Stud gauge and depth generally can be increased without affecting the assembly's fire-resistance rating.

Strength and performance characteristics can be achieved in a variety of ways. Increase wall strength by using materials with a heavier gauge, stronger stud designs, narrower stud spacing, or larger web dimensions. Studs typically are selected to maintain cost control and design integrity. Increased strength requirements generally are met by increasing the metal gauge or stud style, and then, if necessary, increasing the stud dimensions.

Table 2.2 is a basic stud-member selector guide, while Figure 2.6 gives typical steel-stud and runner dimensions. It is important to note that there is a serious misconception within the construction industry regarding the substitution of one manufacturer's studs for

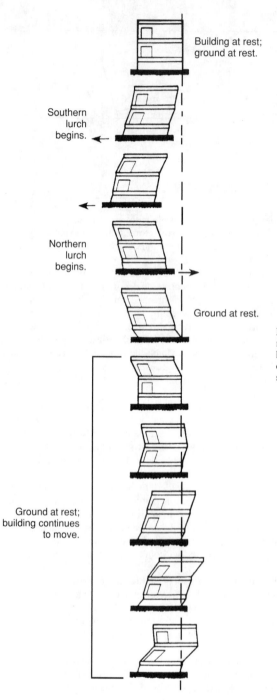

Building at rest;
ground at rest.

Southern
lurch
begins. ←

Northern
lurch
begins.

Ground at rest.

Ground at rest;
building continues
to move.

FIGURE 2.5 When the ground
moves in an exaggerated way, a
house goes through these motions
during and after a simple north-
south lurch.

TABLE 2.1 Gauge Thickness

Nominal gauge[2]	Nominal thickness[3]	Tolerances plus/minus[4]	Minimum thickness[5]	Maximum design thickness[6]
25	0.0209	0.003	0.0179	0.0188
22	0.0299	0.003	0.0269	0.0283
20	0.0359	0.003	0.0329	0.0346
18	0.0478	0.005	0.0428	0.0451
16	0.0598	0.006	0.0538	0.0566
14	0.0747	0.007	0.0677	0.0713
12	0.1046	0.008	0.0966	0.1017
10	0.1345	0.008	0.1180	0.1240

[1]All dimensions are inches, uncoated.

[2]U.S. standard gauge for uncoated hot- and cold-rolled sheets. Gauge numbers are only provided as a reference and should not be used to order, design or specify steel studs or joists.

[3]Nominal thicknesses 0.0209 to 0.0359 are from Table 16 of ASTM A 568 and nominal thicknesses 0.0478 to 0.1345 are from Table 4.

[4]Manufacturing tolerances from ASTM A 568.

[5]Minimum thickness of material delivered to the job site.

[6]Design thickness of steel stud/joists shall not exceed the minimum thickness divided by 0.95. Design thickness = minimum thickness/0.95.

TABLE 2.2 Stud-Member Selector

Flange size	1¼"	1¾"	1⅜"	1⅝"	2"	2½"
Suggested application	Nonload-bearing interior framing	Curtain wall	Curtain wall, light load bearing	Curtain wall, load-bearing joist	Joist-wide flange	Joist-extra wide flange
Gauge range	25–20	20–14	20–14	20–12	20–12	20–12
Available depths*	1⅝"	2½"	2½"	2½"	2½"	3⅝"
	2"	3½"	3½"	3⅝"	3⅝	4"
	2½"	3⅝"	3⅝"	4"	4"	6"
	3"	5½"	4"	5½"	6"	8"
	3½"	4"	5½"	6"	8"	10"
	3⅝"	6"	6"	7¼"	9¼"	12"
	4"	8"	8"	9¼"	10"	13½"
	6"			9½"	12"	
				10"	13½"	
				11½"		
				12"		
				13½"		
				14"		

Typical punch-out or knockout design**

*Availability might vary by geographic region.

**Punch-out or knockout size might vary on smaller web members.

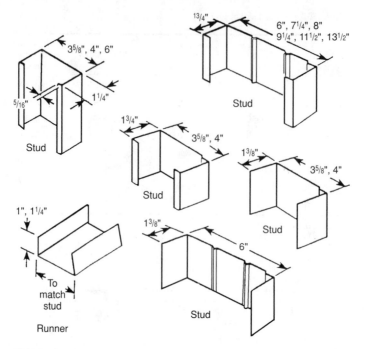

FIGURE 2.6   Steel-framing stud sizes and types.

those of another. The assumption is that all studs of a given size and steel thickness are interchangeable. This assumption is completely false. It is possible that the substitution can be made safely, but the decision should not be made until the *structural* and *physical properties* of the studs involved are compared. Most reliable manufacturers publish structural-property tables in their technical literature. Some light steel-framing studs are ribbed to provide extra vertical strength for load-bearing members. (See Appendix B to learn how to read a typical lightweight steel manufacturer's catalog.)

The steel components shown in Figures 2.7 and 2.8 are lightweight, versatile nonload-bearing members of economical, fire- and sound-barrier systems that can be used in place of masonry for cavity shaft walls, area separation walls, and furring and double-wall compartments in multifamily and multistory lightweight steel dwellings. As described in Chapter 8, exterior curtain walls are nonaxial-load-bearing exterior walls. The steel studs used for curtain-wall applications generally are modified channel-type and roll-formed from five steel thicknesses. The same studs can provide wall framing for both drywall and veneer plaster systems. Exterior surfaces can be brick veneer, portland cement-lime stucco, EIFS, decorative panels, or siding materials.

Placing Punch-Holes

Some studs and joists are manufactured with punch holes in the web to accommodate plumbing and electrical installations (Figure 2.9). For optimum effectiveness, a punch hole is usually provided 12 inches from each end. Intermediate holes are generally placed at 24-inch intervals along the web, with no less than 24 inches between the last intermediate and end hole. (Unpunched studs usually are available on request.)

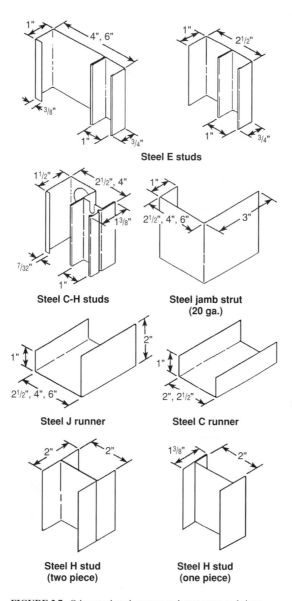

FIGURE 2.7 Other steel stud, runner, and strut types and sizes.

Selecting Other Steel-Framing Components

There are several other steel-framing components that aid in the installation of lightweight steel framing. Descriptions of some of the more common items follow.

Flat Straps. Flat straps are available in various lengths to laterally brace studs and for floor bridging. Strapping also is available in varying widths for diagonal bracing to resist racking (Figure 2.10).

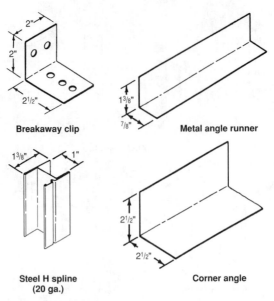

FIGURE 2.8 Steel components used for area-separation walls, shaft walls, and corner-angle applications.

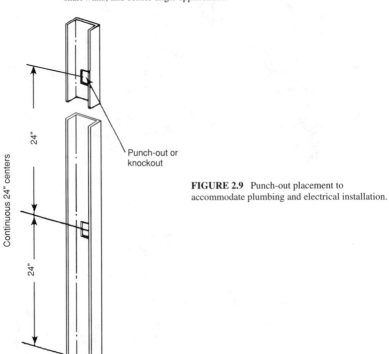

FIGURE 2.9 Punch-out placement to accommodate plumbing and electrical installation.

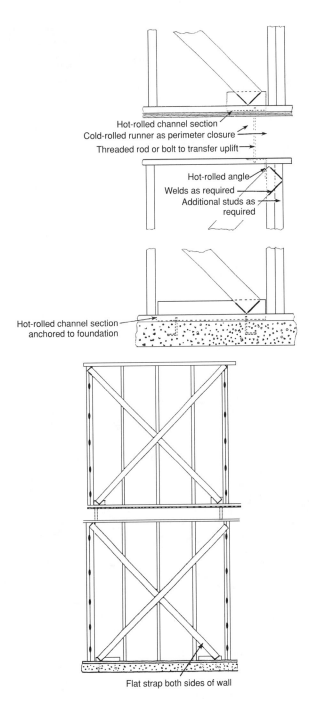

FIGURE 2.10 Diagonal racking flat-strap bracing.

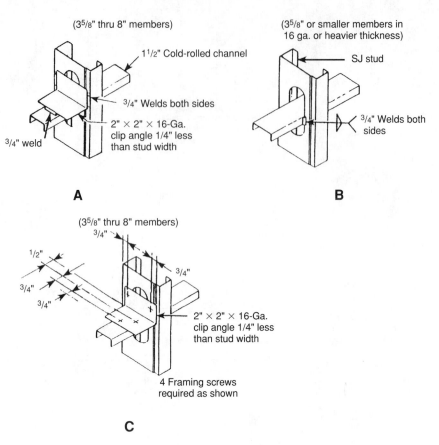

FIGURE 2.11 Methods of attaching lateral bracing: (A and B) weld attachment and (C) screw attachment.

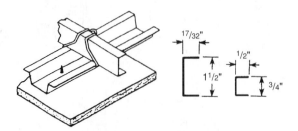

FIGURE 2.12 Typical cold-rolled channel (CRC).

Lateral Braces or Bridging. Properly placed horizontal, lateral braces or bridging provide resistance to stud rotation and minor axis bending under wind and axial loads (Figure 2.11). Attach both stud flanges to the top and bottom runner flanges to provide proper end support. Floor and ceiling runners also must be securely anchored to the structure. To fully utilize the stud's load-carrying capacity, install horizontal bracing at proper intervals.

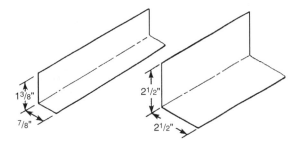

FIGURE 2.13 Angles and angle clips.

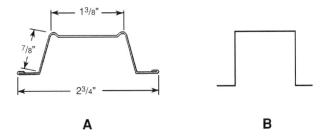

A   **B**

FIGURE 2.14 Drywall furring channel.

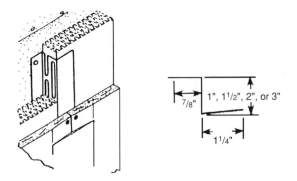

FIGURE 2.15 Z furring channel.

Cold-Rolled Channels. Cold-rolled channels (CRC) facilitate the installation of pipe and conduit and provide optional stiffening. Use channel clips or framing clips to attach the CRC to the studs (Figure 2.12).

Angle Clips. Angle clips provide an economical way to box in a column and perimeter mold for a suspended gypsum ceiling of joists or stud connections (Figure 2.13).

Drywall Furring Channels. Used to attach gypsum board to masonry and steel studs (Figure 2.14), drywall furring channels also can be employed to attach CRC to suspended ceilings.

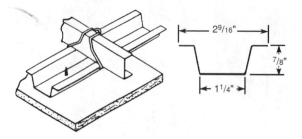

FIGURE 2.16 Metal furring channel.

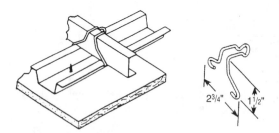

FIGURE 2.17 Furring channel clips.

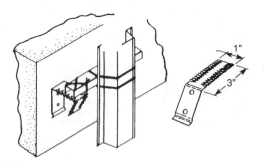

FIGURE 2.18 Adjustable wall-furring brackets.

Z Furring Channels. Z furring channels are used to mechanically attach insulation blankets, polystyrene insulation, and other rigid insulation or gypsum panels or base to the interior side of monolithic concrete or masonry walls (Figure 2.15).

Metal Furring Channels. Metal furring channels (Figure 2.16) are roll-formed, hat-shaped sections that are designed to attach wallboard panels and gypsum base to wall and ceiling furring using screws.

Furring Channel Clips. Furring channel clips are used to attach metal furring channels to 1½-inch CRC ceiling grillwork (Figure 2.17). Designed to be used with single-layer gypsum panels or base, the clips are installed on alternate sides of the 1½-inch channel. Where the clips cannot be alternated, wire tying is recommended.

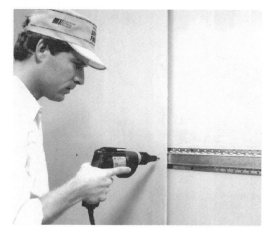

FIGURE 2.19 Resilient furring channel (RC-1).

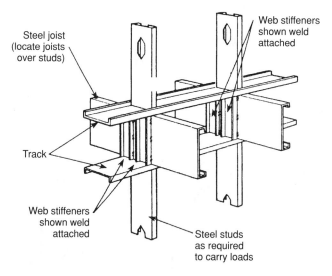

Steel joist (locate joists over studs)

Web stiffeners shown weld attached

Track

Web stiffeners shown weld attached

Steel studs as required to carry loads

FIGURE 2.20 Typical web-stiffener assembly.

Adjustable Wall-Furring Brackets. Adjustable wall-furring brackets (Figure 2.18) are used to attach CRCs and metal furring channels to the interior side of exterior masonry walls.

Resilient Furring Channels. Resilient furring channels (RC-1) help control sound transmission through steel-framed walls and ceilings (Figure 2.19).

Web Stiffeners. Web stiffeners attach either inside or outside of the joist web when required to increase load-carrying capacity and prevent web crippling at points of reaction and concentrations of axial loads (Figure 2.20).

Wall Slip Tracks. Wall slip tracks might be required to accommodate the deflection of floor beams or floor decks above curtain-wall or interior partitions (Figure 2.21). Slip tracks cannot be used in axial-load-bearing stud conditions or above continuous window spandrels.

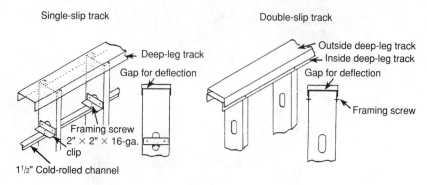

FIGURE 2.21 Wall slip track.

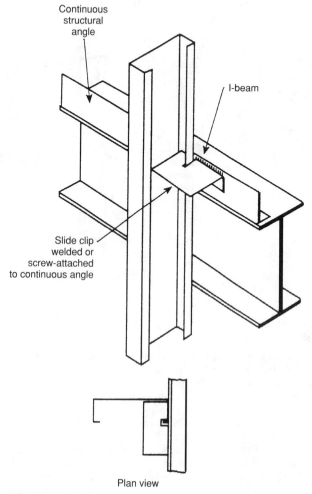

Plan view

FIGURE 2.22 Wall slide-clip arrangement.

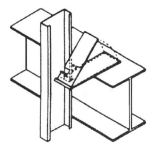

FIGURE 2.23 Lovely wall slip anchor.

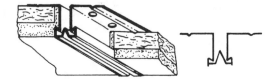

FIGURE 2.24 Typical control joint.

Wall Slide Clips. Wall slide clips generally are used in curtain-wall applications to attach horizontal supports. They allow vertical structural movement without transferring compressive loads to the studs (Figure 2.22). The curtain-wall slide clip resists tension loads caused by *negative* wind pressure, or suction, on exterior curtain walls.

Lovely Wall Slip Anchors. Lovely wall slip anchors (Figure 2.23) often facilitate the attachment of the curtain wall to the horizontal structural support while allowing the structure to move vertically.

End Clips. End clips attach the ends of joists, headers, and other light steel-framing members that would otherwise be unsupported.

Foundation Clips. Foundation clips are 16-gauge angles designed to secure parallel joists or C-runner end caps to the foundation.

Control Joints. Control joints (Figure 2.24) are used to relieve stresses caused by expansion and contraction across the control joint in large ceiling and wall expanses in drywall and veneer-plaster systems. They are used from door header to ceiling, from floor to ceiling in long partitions and wall furring runs, and from wall to wall in large ceiling areas. Made from roll-formed zinc to resist corrosion, the joints have a ¼-inch open slot, which is protected by plastic tape that is removed after finishing. Ceiling-height door frames can be used as control joints. For less than ceiling-height door frames extend control joints to the ceiling from both corners.

Other lightweight steel-framing metal accessories include Z purlins, J tracks, L trims, casing beads, and trims.

All light steel-framing members and accessories should meet ASTM A568 for uncoated steel thickness; ASTM A446 grade A (33-ksi yield strength), grade C (40 ksi), or grade D (50 ksi) for stud-joist members; and grade A (33 ksi) for cold-rolled members. Metal building components usually are produced from galvanized steel that meets ASTM A525 standards.

USING TOOLS OF THE TRADE

Only a few essential tools are required to install lightweight steel framing and perform two basic operations: cutting and fastening (Figure 2.25).

FIGURE 2.25 Essential tools for cutting and fastening steel.

FIGURE 2.26 The chopsaw is the most used and efficient steel-cutting tool.

FIGURE 2.27 Typical metal-cutting saw blade.

FIGURE 2.28 Circular saw equipped with a reinforced abrasive cutoff wheel.

Cutting Steel

Cold-formed steel-framing components can be ordered cut to length. Some onsite cutting, however, is necessary for most projects. The most efficient means of cutting steel is with a chopsaw (Figure 2.26) fitted with either a friction-type blade, which has 280 to 300 teeth, or a reinforced abrasive cut-off wheel. If a metal cutting blade is used, it should have a minimum diameter of 14 inches and a thickness of ⅛ inch (Figure 2.27). The saw should have a minimum of 5 horsepower (hp) in order to efficiently cut the steel. Its base can be placed on a bench, sawhorse, or floor to quickly and efficiently gang-cut pieces.

Hand cut steel studs, runners, and joists with a 3-hp, worm-drive, handheld circular saw equipped with a reinforced abrasive cut-off wheel (Figure 2.28). A cut-off saw (Figure 2.29) with an abrasive wheel provides more power than a circular saw. For onsite work, the cut-off saw also can be used to cut heavier-gauge steel members. Power bandsaws and portable power hacksaws (Figure 2.30) also might be used, but they usually are not efficient on large projects.

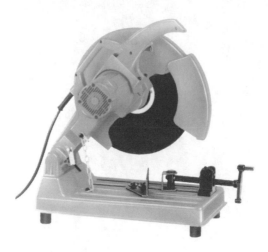

FIGURE 2.29 Typical cutoff saw.

Exercise care when using all cutting equipment. Shield blades to prevent accidents, contain sparks, and protect the worker in case the blade fractures. Never torch-cut a load-bearing steel member.

Descriptions follow of some of the other tools that a steel framer should have available.

Metal Snips. Metal snips make straight and curved cuts in steel-framing components (Figure 2.31A). Snips are available in several sizes and styles. Offset snips, shown in

A

FIGURE 2.30 Use (A) bench-type power bandsaw or (B) portable power hacksaw cut lightweight metal framing components.

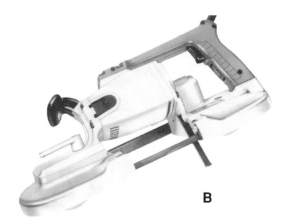

FIGURE 2.30 *(Continued).*

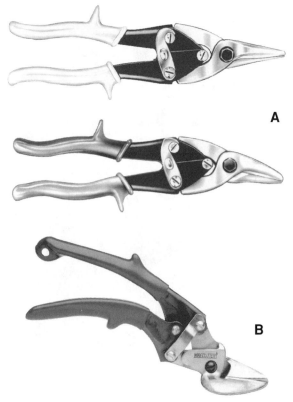

FIGURE 2.31 (A) Straight and curved metal snips; (B) offset metal snips.

FIGURE 2.32 The steel framer has many uses for end nippers.

FIGURE 2.33 Channel-stud shear.

FIGURE 2.34 Electric shears can cut up to 20-gauge steel.

FIGURE 2.35 Pneumatic-type shears are more powerful than electric shears.

Figure 2.31B, keep hands safely above the work and cut up to 18-gauge material. Do not attempt to cut heavier materials than those for which the snips are designed.

Lather's Nippers. Lather's, or endcut, nippers (Figure 2.32) are ideal for making and cutting wire-tied attachments to framing components.

Channel-Stud Shears. This tool cuts steel studs and runners quickly and cleanly without deforming them (Figure 2.33). Most stud shears have fixed guides for 1⅝-, 2½- and 3⅝-inch sizes. Usually they cut a maximum steel thickness of 20 gauge.

Electric Power Shears. Electric power shears (Figure 2.34) cut metal-framing studs, runners, steel metal, and cold-rolled steel up to 20 gauge. The cutting head on most shears rotates 180 degrees for side and overhead work. Cutters are replaceable. Pneumatic power shears also are available (Figure 2.35).

Power Nibblers. A power nibbler can be used to make quick cuts in up to 16-gauge steel (Figure 2.36).

FIGURE 2.36 Nibblers are good for making quick metal cuts.

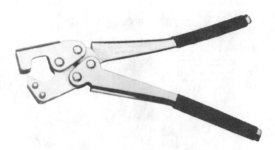

FIGURE 2.37   Stud crimper tool.

FIGURE 2.38   Typical air compressor.

Stud Crimpers.   This tool is used to set and splice steel studs, rough-in door holders and window headers, set electrical boxes, and punch out hanger-wire holes in ceiling grids (Figure 2.37).

Air Compressors.   An air compressor is needed to operate air, or pneumatic, tools. The unit shown in Figure 2.38 can operate three screw guns simultaneously.

Fastening Materials

Residential steel-framing studs, runners, and joists can be fastened together with screws or welds. The actual fastening tools and their methods of use are described in Chapter 3.

Combination Chalkline Box and Plumb Bob.   The plumb-bob-shaped device holds a 100-foot chalkline and chalk.

Clamps.   Clamps are valuable third hands for the steel framer. There are several types available, but the locking C clamp is still the most popular (Figure 2.39A). Special clamps, such as the retrofilled vise-grip clamp shown in Figure 2.39B, can help tighten X-bracing straps.

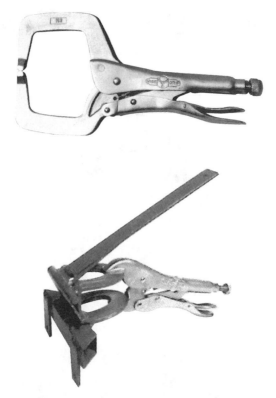

FIGURE 2.39   Two types of useful clamps.

Aligning Studs and Runners

The plumb bob, carpenter level, and tape measure commonly are used in wood-framed installations. A tape measure is used to establish multiple points on the floor in order to lay out the walls. A chalkline was popped along these points to establish the wall location. The 3-4-5 measurement method was used to lay out 90-degree angles for partitions.

These tasks, which were done on the floor, were relatively easy, even though they required two people. The hard part was transferring all the measurements to the ceiling or top of the wall. The tools used to accomplish this were either the plumb bob or carpenter level. The plumb bob requires two people and is less than desirable to work with at times: Ever try to use even a 16-ounce plumb bob on a windy day?

The carpenter's level, which is used to plumb studs, works well on standard 8-foot-high walls (Figure 2.40), but cannot be used on tall walls. Higher walls have to be plumbed with the plumb bob, and the higher the wall, the more difficult it is to get the plumb bob stationary. While these tools can be used to lay out steel studs and runners, more and more contractors are employing a laser (Figure 2.41).

All a laser needs is a limited number of control points, which can be established with a tape measure. Generally, the control point to plumb interior studs is within the building. If control is to be performed from the outside, position the laser on its side on a tripod or di-

FIGURE 2.40 Carpenter's level.

FIGURE 2.41 The most useful type of laser-adjustment tool for the steel framer.

FIGURE 2.42 Use a laser to plumb wall studs.

FIGURE 2.43 Typical handheld laser.

rectly on the ground at a known offset. Position the beam over one point and use an electronic receiver and the laser's line control to position the rotating vertical beam to the far point. After setup is complete, the beam rotates on a vertically plumb beam at a known offset and can be used to lay out and plumb the studs (Figure 2.42). An electronic receiver is attached to a 2-×-2-inch rod and placed at the proper distance to establish the line against which the studs are plumbed.

The handheld laser measuring tool shown in Figure 2.43 measures effectively from 10 inches up to 330 feet with accuracy to within ±⅛ inch. The digital display gives readings in either metric or foot units. To get an accurate measurement, point the visible laser dot to the distant target and read the distance on the digital display.

Recommending Basic Tools

Earlier in this chapter, descriptions were given for essential or desirable steel-framing tools for the individual framer. For the steel-framing contractor, Table 2.3 provides a recommended basic tool package for a crew with four to five workers.

TABLE 2.3 Required Tools for Steel-Framing Crew

Quantity	Tools — Item
4	Assorted metal files
4–5	Belts with bolt bags
1	Brush, wire
4–6	6-inch C clamps vise grips
2	Retro-filled clamps
2	Caulking guns (open barrel)
2	100-foot chalklines with chalk
1	Concrete chisel

TABLE 2.3 Required Tools for Steel-Framing Crew (Continued)

	Tools
Quantity	Item
2	Hammers, carpenter's straight claw
1	Hammer, sledge (10 pound) long handle
1	Hammer, sledge (3 pound) short handle
2	Hammers, soft face
4–5	Hard hats
4–5	Knives, utility (retractable)
1	Plumb bob
3	Snips, aviation (1 right hand, 1 left hand, and 1 straight-offset preferred)
1	Spirit level, 4 to 6 foot, magnetic
1	Spirit level, 2 foot, magnetic
1	Square, framing
4	Square, speed
1	Laser-alignment tool
6	Carpenter pencils
2	Bright thin-line marker (yellow or blue)
4–5	Tape, measuring 25–30 feet, steel
1	Tape, measuring 100 feet, steel wrenches
2 each	Box/open combination ($^{15}/_{16}$, $1^1/_{16}$, and $1^1/_4$ inch)
2	Spud wrenches, $1^1/_4$ inch
2–3	Ratchets, $^1/_2$-inch drive
2–3	Sockets, $1^1/_4$-inch, $^1/_2$-inch drive
2	Sockets, $1^1/_{16}$-inch, $^1/_2$-inch drive
2	Sockets, $^{15}/_{16}$-inch, deep, $^1/_2$-inch drive
1	Channel stud shear
1	Stud crimper
2	End cut nippers
1	Metal stud punch and Grommet KitPower tools
3	Power nibbler
2	Power shears
1	Rotary hammer drill ($2^1/_4$ by 6- to 8-inch-long masonry drill bit)
6	$^5/_{16}$-inch magnetic hex nut drivers
1	Powder-actuated tool kit with $1^1/_4$-inch, screw and yellow shots
4	Screw guns, 2500 rpm variable speed, $4^1/_2$- to 5-amp with depth
2	$^1/_4$-inch magnetic hex nut drivers
6	$^5/_{16}$-inch hex magnetic nut drivers
6	Phillips tips, No. 2
6	Tip holder, magnetic

TABLE 2.3 Required Tools for Steel-Framing Crew (Continued)

Quantity	Item
	Tools
1	14-inch heavy duty chopsaw with base and metal cutting blades
1	⅜-inch circular saw
1	Air compressor equipment—buy or rent
2	8-foot step ladder, fiberglass
1	Extension ladder, 22 to 28 feet
2	Section minimum of rolling scaffolds, 6 feet each
1	All terrain forklift
1	Crane, 70-foot boom reach minimum materials handling
2–4	⅜-inch-diameter chokers with eyes both ends—10 feet long
6	½-inch-diameter rope—50 feet long with hooks
4	Cable pullers—1-ton rating
4	Come-along or manual hoist
	Suggested miscellaneous
1–3	Brooms
1–2	Buckets
As required	Extension cords
1	Ground fault interrupt
As required	Fire extinguishers
As required	First aid kit
1–25	Gallon water cooler with cups
1	Flashlight
10	Pair work gloves
1	Shovel
As required	Sponges and shop rags
1	Roll nylon string
2–4	Safety belts (harness) for overhead work

PRACTICING SAFETY PROCEDURES

Safety is very important to the steel framer. Workers who are inexperienced with any of the tools or equipment, the described procedures, or are uncertain about the safety of the procedures for a particular activity or undertaking, must consult with someone who is skilled or certified in this work area before beginning a task. Safeguard against possible injury by seeking help from a more experienced worker or supervisor when uncertain about proper safety measures. Safety procedures can preserve time, material, property, and equipment.

Safe work practices can make the job easier and more enjoyable, as well as help prevent harmful and costly accidents. Remember and follow these safety tips when using steel-framing tools.

- Keep tools in good working order at all times and regularly inspect tools under a specified maintenance program.
- Only allow authorized personnel to use tools and then only for the purpose for which the tool was designed.
- All power tools should have a three-wire ground system. Carefully inspect each tool to make sure the case is grounded.
- Frequently inspect extension cords and connectors and keep them in good condition. Insulation and protective coating must remain intact. The recommended extension cord size for 4-amp load is as follows:

Length in feet	0–50	50–100	100–150	150–200
Wire gauge	18	16	14	12

- Extension cords should be 3-wire cords with twist/lock connections, UL approved, and in compliance with the Occupational Safety and Health Administration (OSHA).
- Locate electrical switch boxes in an easily accessible area and away from possible hazards such as water, unauthorized personnel, and moving vehicles. Permanently attach a locking device to the switch box and lock it at the end of each work day.
- Circuits should be properly fused.
- Properly train workers in the use of the equipment.
- When drilling, grinding, or sawing, operators should wear goggles, and use guards and other safety devices.
- Workers should be able to operate power tools without stretching too far or standing in an unbalanced position.
- Under no circumstances should a worker use power tools while standing on wet ground, in the rain, or when wearing wet gloves or shoes.
- In the event that the operator does get a shock, he or she should not attempt to grab the defective tool with both hands.
- Regularly inspect powder-actuated tools and keep them in proper working order. Authorize only responsible, trained, and licensed personnel to use powder-actuated tools.
- Never leave a powder-actuated tool unattended when loaded.
- Forbid carelessness and horseplay.

USING SCAFFOLDING AND LADDERS

Portable scaffolding is ideal for jobs that do not require full scaffolding (Figure 2.44). Although easy to set up, scaffolding and ladders can be dangerous if care is not taken. To be sure of safety when using these two tools follow these guidelines.

- Erect or construct scaffolding and ladders in a safe, suitable, and proper manner. Provide solid footing and secure anchoring.
- Erect scaffolding so it does not present an obstruction or danger to personnel below.

FIGURE 2.44 Two typical scaffolds.

- When using rolling scaffolding, lock the wheels prior to mounting the scaffold. All scaffolding of this nature should have handrails and a full deck.
- When the scaffolding cannot be placed on a flat surface, build a level base underneath the scaffolding that will not shift and cause the scaffold to fall.

- If the scaffolding is more than two sections high, tie it off.
- The scaffolding decking lumber should be knot free and periodically inspected for defects.
- Properly make and inspect all scaffolding connections before climbing onto the section. Never roll the scaffolding when someone is on it.

CHAPTER 3
FASTENERS FOR RESIDENTIAL STEEL FRAMING

The framing components of residential, cold-formed steel-framed systems, as in wood-framed systems, include studs, joists, and rafters or trusses. These vertical and horizontal framing members commonly serve as structural load-carrying components and utilize durable and dependable connectors that develop positive connections, primarily with screws and welds. These steel systems require fastening equipment and labor skills similar to those used for wood-framed construction.

The fastening equipment enables fast, economical, and versatile connections. Selecting a fastener for a particular job depends on load conditions; thickness, strength, and configuration of the materials; the contractor's experience; and the availability of fasteners. Finding the best fastener for these factors is the key to an economical and efficient steel-framing system. Keep in mind that steel framing uses an assortment of off-the-shelf fasteners. The following discussion provides insight into the development of an efficient fastening system by providing guidance on the use and design of fasteners for cold-formed steel framing.

SELECTING SCREWS

The most common steel-framing fastener is the self-drilling screw. In one operation, it can drill the hole and securely fasten just about any material to steel framing. These screws come in a variety of styles to fit a vast range of requirements (Table 3.1). For exterior applications, screws are available with zinc, cadmium, or copolymer coatings.

Describing Screwhead Styles

Self-drilling screws are manufactured in a variety of head configurations to meet specific installation needs (Figure 3.1). The most common driving recess for the screwhead is the No. 2 Phillips design, but others also are available.

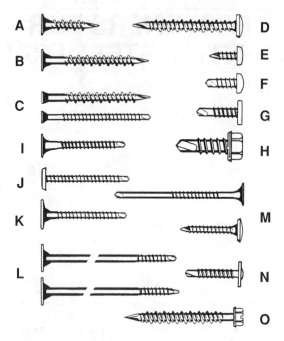

FIGURE 3.1 Various self-drilling screws available to the steel framer. Screws marked A, B, C, I, K, L are buglehead screws; D is a type-S ovalhead; E and F are type-S panhead; G is a type-S-12 low-profile head; H is a hex-washer head; J is a type-S-12 pancakehead; M is a type-S-12 pilot-point; N is a modified trusshead; and O is an acorn-slotted hexhead.

Buglehead. This head is designed to countersink slightly in the gypsum wallboard or sheathing, plywood, or finishing material without crushing the material or tearing the surface. It leaves a flat, smooth surface for easy finishing.

Low-Profile Head. The low-profile head maintains a pleasing appearance at fastening points when the application does not require heads to be flush with the surface. Use this style when rigid sheathing or finishing materials are installed on top of the screwhead.

Panhead. This common head configuration generally fastens studs to track; connects steel bridging, strapping, or furring channels to studs or joists; and steel door frames to studs.

Hex-Washer Head. This is a common screwhead for penetrating steel and usually is used on thicker steel materials. The washer face provides a bearing surface for the driver socket, which assures greater stability during driving. The ¼-inch-size head is most common. A ⁵⁄₁₆-inch head usually is required for heavy applications.

NOTE: Consider using an alternate lower-profile head when applying rigid finishing materials across the top of these screws.

Ovalhead. Use this head when the accessory that is to be attached to the framing has oversized holes or requires a low-profile appearance. This style of head is used to attach cabinets and brackets to framing.

Trimhead. Used for fastening wooden trim and other thick or dense finishing materials to steel studs, the small trimhead sinks into the trim material. The head allows for easy finishing with minimal disturbance of the material surface.

Waferhead. Larger than the flathead and buglehead screws, the waferhead is used to connect soft materials to steel studs. The large head provides an ample bearing surface yet sits flush to achieve a clean, finished appearance.

Trusshead. The trusshead is an extremely low-profile head that is commonly used to attach metal lath to steel framing (also referred to as lathhead). Use this head to attach metal lath to steel framing up to 14 gauge.

Flathead. Designed to countersink and sit flush without causing wooden flooring or finishes to splinter or split (not illustrated).

TABLE 3.1 Framing Screw Selector

Fastening application	Fastener used	Figure 3.1
	Panels to steel framing	
½" single-layer panels to steel studs, runners, channels	1" type-S buglehead	A
⅝" single-layer panels to steel studs, runners, channels	1" type-S buglehead 1⅛" type-S buglehead	A A
¾" single-layer panels to steel studs, runners, channels	1¼" type-S buglehead	A
1" coreboard to metal angle runners in solid partitions	1⅝" type-S buglehead	A
½" double-layer panels to steel studs, runners, channels	1⅝" type-S buglehead	B
⅝" double-layer panels to steel studs, runners, channels	1⅝" type-S buglehead	B
¾" double-layer panels to steel studs, runners, channels	2¼" type-S buglehead	B
½" panels through coreboard to metal angle runners in solid partitions	1⅞" type-S buglehead	B
⅝" panels through coreboard to metal angle runners in solid partitions	2¼" type-S buglehead 3" type-S buglehead	B B
1" double-layer coreboard to steel studs, runners	2⅝" type-S buglehead	B
	Wood to steel framing	
Wooden trim over single-layer panels to steel studs, runners	1" type-S trimhead 1⅝" type-S trimhead	C C
Wooden trim over double-layer panels to steel studs, runners	2¼" type-S trimhead	C
Steel cabinets, brackets through single-layer panels to steel studs	1¼" type-S ovalhead	D
Wooden cabinets through single-layer panels to steel studs	1⅝" type-S ovalhead	D
Wooden cabinets through double-layer panels to steel studs	2¼", 2⅞", 3¾", type-S ovalhead	D

TABLE 3.1 Framing Screw Selector (Continued)

Fastening application	Fastener used	Figure 3.1
Steel studs to door frames, runners		
Steel studs to runners	⅜" type-S panhead	E
Steel studs to door frame jamb anchors	⅜" type-S-12 panhead	F
Other metal-to-metal attachment (12-ga. max.)	⅝" type-S-12 low-profile head	G
Steel studs to door frame jamb anchors (heavier shank ensures entry in anchors of hard steel)	½" type-S-12 panhead ⅝" type-S-12 low-profile head	F G
Strut studs to door frame anchors, rails, other attachments in movable partitions	½" type-S-16 panhead with anticorrosive coating	F
Metal-to-metal connections up to double thickness of 12-ga. steel	¾" S-4 hex washer head with anticorrosive coating	H
Gypsum panels to 12-ga. (max.) steel framing		
½" and ⅝" panels and gypsum sheathing to steel studs and runners; specify screws with anti-corrosive coating for exterior curtain-wall applications	1" type-S-12 buglehead	C
Self-furring metal lath and brick wall ties through gypsum sheathing to steel studs and runners; specify screws with anticorrosive coating for exterior curtain-wall applications	1¼" type-S-12 buglehead 1¼" type-S-12 pancakehead	I J
½" and ⅝" double-layer gypsum panels to steel studs and runners	1⅝" type-S-12 buglehead	I
Multilayer gypsum panels and other materials to steel studs and runners	1⅞", 2", 2⅜", 3" type-S-12 buglehead	I
Cement board to steel framing		
Cement board or exterior cement board direct to steel studs, runners	1¼", 1⅝" steel screws	K
Rigid foam insulation to steel framing		
Rigid foam insulation panels to steel studs and runners for 20-, 25-ga. steel	1½", 2", 2½", 3", type-S-12	L
Aluminum trim to steel framing		
Trim and door hinges to steel studs and runners (screw matches hardware and trim)	⅞" type-S-18 ovalhead with anticorrosive coating	M
Batten strips to steel studs in demountable partitions	1⅛" type-S buglehead	A
Aluminum trim to steel framing in demountable and partitions	1¼" type-S buglehead with anticorrosive coating	A

TABLE 3.1 Framing Screw Selector (Continued)

Fastening application	Fastener used	Figure 3.1
Plywood to steel joists		
⅜" to ¾" plywood to steel joists (penetrates double thickness 14-ga.)	1¹⁵⁄₁₆" type-S-12 buglehead, pilot point	M
Steel to poured concrete or block		
Attachment of steel framing components to poured concrete and concrete block surfaces	³⁄₁₆" × 1¾" acorn-slotted anchor	N
Metal lath to steel framing		
Attaches metal lath to steel framing up to 14-ga.	½" to 1¼", trusshead	O

Choosing a Screw Point

Two common screw points are used in cold-formed steel framing: type-S and type-S-12 (Figure 3.2). These types generally refer to ASTMC-1002- and ASTMC-954-type screws, respectively. The American Society for Testing and Materials (ASTM) designations refer to specifications for steel screws for the application of gypsum board or metal plaster bases to steel studs. Type-S is for steel up to 0.035 inch thick. The point might be a sharp needle or piercing point similar to that used on wood screws, or it might be fluted to aid in the drilling process. Type-S-12 is designed for steel up to 0.112 inch thick and has a shorter, fluted tip. For thicker materials, contact the manufacturer for the proper point type.

Depending on the thickness of the material, the thread along the screw's shank should be held back from the point of the screw to prevent the threads from engaging the steel until the drilling process is complete. This prevents the overdrilling of the first ply or stripping the threads after partial penetration. It also tightly pulls the plies together.

If you are connecting wood or other rigid material to steel, the threads should not continue to the screwhead. Instead the threads should allow the screw to draw the plies together with a minimal uplift of the wood or rigid material. The choice of screw points and thread configurations depends on how the fastener is required to perform the installation. Test various configurations to find the type that performs most efficiently with different materials.

Deciding on Screw Diameters

Screws for steel framing are available in the following sizes: #6, #7, #8, #9, #10, #11, #12, #13, and #14. The smallest diameter screw is #6. Specify sizes based on the required ca-

FIGURE 3.2 Common screw points (left to right): type-S needle-point, type-S fluted needle-point, and type-S-12 fluted-point.

TABLE 3.2 Screw Diameter Size Guidelines
Based on Total Thickness of Steel

Common size	Nominal diameter		Total thickness (in.)
#6	0.138		0.110
#7, #8	0.151, 0.164	Up to	0.140
#10	0.190		0.175
#12	0.216		0.210*
¼	0.250		0.210*

*Greater thicknesses are possible; consult a screw manufacturer.

pacity of the connection, the length of the screw, and the thickness of the steel. The most common sizes are from #6 to #10. Table 3.2 provides some guidance on the diameters typically required for various steel thicknesses.

Selecting a Screw Length

Generally, screws should be approximately ⅜ to ½ inch longer than the thickness of the connected materials. At least three exposed threads should extend through the steel to assure an adequate connection. These are minimum guidelines. It is important to select the correct fastener for each application. Consult the manufacturer's specifications for the correct screw type and length.

Table 3.3 summarizes the various lengths of screws used for different material thicknesses. The sizes shown are those commonly used in cold-formed steel framing, but the table by no means covers all available sizes. Manufacturers can provide a more complete list of available products.

TABLE 3.3 Common Screw Lengths

Common sizes (#)	Length	Materials and thickness of materials that can be connected
6, 7	⅜"–7⁄16"	Steel to steel
6, 8, 10, 12	½"	Steel to steel
8, 10	⅝"–1"	Steel to steel; metal lath to steel
10, 12, 14	¾"–1½"	Steel to steel
6, 8	1"	½" or ⅝" panels to steel
8, 10, 12	1¼"	Self-furring metal lath or masonry ties to steel
10	1½"	Temp., panel-panel lamination
6, 7	1½"–2"	Multiple layers to steel
6, 7	2¼"	Wooden trim and 3-layers of panels to steel
8	2"–3½"	Multiple layers to steel
10	3½"–6"	Multiple layers to steel

Determining Load Capacities and Spacing Requirements

Maintain a minimum dimension of 3× the nominal screw diameter (d) to the edge of the steel and to the center of other fasteners for steel-to-steel connections. The edge distance can be reduced to 1.5d in the direction perpendicular to the load (Table 3.4). Connectors for gypsum wallboard and plywood should not be spaced less than ⅜ inch from the edges or ends of the board.

Additional fastener spacing criteria for gypsum board and plywood can be found in the Gypsum Association's (GA) specification (GA-216) and the American Plywood Association's (APA) *Design and Construction Guide/Residential and Commercial* (E-30), respectively. Contact the manufacturer for fastener spacing requirements for other materials.

In the absence of the specification, the following values can be used for the allowable shearing force per inch of fillet or groove welds.

Steel gauge	Design thickness	Weld size	Allowable shear (pounds/inch)
12	0.1046	⁵⁄₃₂	1035
14	0.0747	⅛	740
16	0.0598	⅛	592
18	0.0478	⅛	473
20	0.0359	⅛	355

NOTE: These values can be increased by ⅓ for wind or seismic loading. When joining members of different thicknesses, use value shown for the thinner member.

Applying Screws

Metal framing screws are generally applied with a positive-clutch electric power tool, commonly called an *electric screwgun*, equipped with an adjustable screw-depth control head and a Phillips bit (Figure 3.3). Screws positively and mechanically attach the metal base to the steel framing.

TABLE 3.4 Suggested Capacity for Screws Connecting Steel to Steel (in Pounds)

Steel thickness- thinnest component	¼–14 screw		#12–14 screw		#10–16 screw		#8–18 screw		#6 shear	
	Shear	Pullout	Shear	Pullout	Shear	Pullout	Shear	Pullout	Shear	Pullout
0.1017"	1000	320	890	280	780	245	675	210	560	175
0.0713"	600	225	555	195	520	170	470	145	395	125
0.0566"	420	180	390	155	370	135	340	115	310	95
0.0451"	300	140	280	120	260	105	240	90	220	75
0.0347"	200	110	185	95	175	80	165	70	150	60

FIGURE 3.3 Typical adjustable electrical screwgun.

NOTE: Do not use a drywall screwgun. They are rated at 4000 revolutions per minute (rpm). At this speed, the self-drilling tip generally burns up before the drilling operation is complete. The correct tool is a variable speed 0-to-2500-rpm screwgun rated at 4½ amps and fitted with a depth-locating nosepiece.

Follow these guidelines for the proper use of a single screwgun.

Adjusting the Screwgun.   Set the adjustment for the proper screw depth. For most situations, set the screwhead flush with the base surface. To adjust the depth, rotate the control head to provide proper screw depth. When properly adjusted, secure the control head to maintain the adjustment (Figure 3.4A).

Placing Screws.   The Phillips-head tip holds the screw for driving (Figure 3.4B). The bit tip does not rotate until pressure is applied to the metal base during application.

Starting Straight Screws.   Maintain a firm grip on the electric screwgun to achieve a straight line of entry. To avoid stress on the wrist, hold the gun as shown in Figure 3.4C and not by the pistol grip. The screw must enter perpendicular to the base face for proper performance. Drive the screws at least ⅜ inch from ends or edges of base.

Operate the electric screwgun constantly. When the screwhead is driven solidly against the base, the screwgun head automatically stops turning as the positive clutch disengages. Exert adequate pressure to engage the clutch and prevent *walking*. A one-piece socket makes driving easier and more efficient than separate socket and extension pieces because it provides for more rigid and firmer control.

The bit tip holds the screw it has driven (Figure 3.5). Depth finders can help ensure proper penetration (Figure 3.6). Use a clamp, usually a C clamp, to hold the steel members in place. Clamping during fastening helps assure that the members are in full contact and prevents overdrilling the hole as the plies pull together. When connecting steel materials of different thicknesses, make the connection from the thinner material to the thicker, so the screwhead brings the thinner, more flexible material in tight contact with the heavier material.

Figure 3.7 shows an automatic-feed screwgun system that adapts to most guns and is 3 to 5 times faster than the conventional application method. This device handles a wide

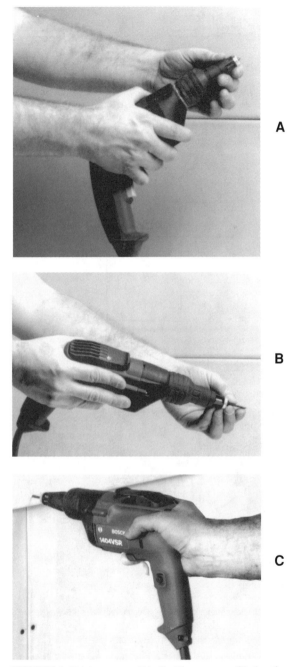

FIGURE 3.4 Driving a screw: (A) adjust the screwgun; (B) place the screw in bit tip; and (C) start the screw straight.

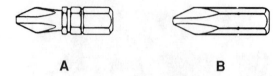

A   **B**

FIGURE 3.5 Two screwgun bit types: (A) No. 1 bit used to drive trimhead and pancakehead screws and (B) No. 2 used to drive buglehead, panhead, low-profile, and ovalhead screws.

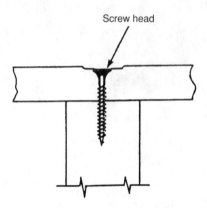

FIGURE 3.6 Proper depth of a driven screw.

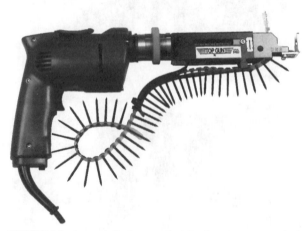

FIGURE 3.7 Automatic-feed screwgun that adapts to most standard guns.

variety of screws. Another power metal screwgun is a 12-volt, keyless chuck, hi-torque, cordless ⅜-inch model (Figure 3.8). The cordless models are efficient for the many situations when electricity is not readily available or when it is difficult to run an extension cord.

FIGURE 3.8 Typical keyless chuck hi-torque cordless 3/8-inch driver/drill.

FASTENING WITH DRIVE PINS AND NAILS

Drive pins and some types of nails can be used to fasten steel frames. The holding strength of these pins depends on the shank diameter, depth of penetration, spacing, and edge condition.

Using Pneumatically Driven Fasteners

Pneumatically, or air driven, fasteners are fairly new to the cold-formed steel-framing industry. This equipment uses techniques similar to the nailguns (Figure 3.9) used for wooden framing. Pneumatic fasteners commonly are used to nail plywood to steel because

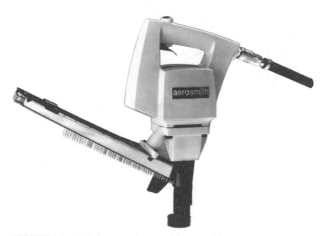

FIGURE 3.9 Typical pneumatic automatic screwdriver.

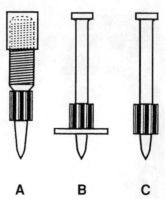

FIGURE 3.10 Three spiral shank nails or pins: (A) threaded head, (B) threaded head with washer, and (C) standard headed pins.

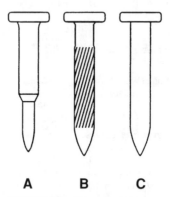

FIGURE 3.11 Other pins: (A) step-down, (B) knurled, and (C) smooth.

of the holding power of their threaded heads with washers (Figure 3.10). These fasteners also are available in standard-head designs to fasten thicker cold-formed steel, concrete, or solid concrete block. Pneumatically driven fasteners now are being developed for steel products less than 0.056 inch thick.

Other fasteners are available with mechanical or electro-zinc plating, or copolymer coatings, depending on the corrosion-resistance requirements. Head diameters range from ¼ to ⅜ inch, shank diameters vary from 0.100 to 0.236 inch, and lengths range from ½ to 8 inches. The shank styles are smooth, knurled, or stepdown (Figure 3.11). Stepdown shanks generally are used on thicker materials for greater load-carrying capacities. Pins typically are supplied in bulk, collated strips, or coils, depending on tool requirements. Tables 3.5 and 3.6 contain suggested capacities for fasteners in concrete and steel. Contact the pin manufacturer for information on the load-carrying capabilities of specific fasteners.

Employing Powder-Actuated Fasteners

Cold-formed steel-framing members can be fastened to concrete or to structural steel with headed drive pins or threaded studs driven by powder-actuated tools. Powder-actuated fas-

TABLE 3.5 Suggested Capacity for Power-Driven Fasteners Connecting Cold-Formed Steel to Structural Steel (in Pounds)

Sheet-steel thickness	Shank diameter: 0.145"		Shank diameter: 0.177"		Shank diameter: 0.205"	
	Structural steel thickness		Structural steel thickness		Structural steel thickness	
	¼"	⅜"	¼"	⅜"	¼"	⅜"
0.1046"	210	210	335	395	485	525
0.0740"	210	210	335	395	485	525
0.0592"	210	210	335	395	465	465
0.0478"	210	210	321	321	372	372
0.0359"	197	197	241	241	279	279

NOTES:

Fastener should project completely through the structural steel component.

Fasteners should not be located less than ½" from the edge of the steel. Minimum spacing of fasteners should be 1½".

Capacity values are for shear or pullout for cold-formed steel with Fy-33ksi. One-third increase for wind should not be included.

TABLE 3.6 Suggested Capacity for Power-Driven Fasteners in Concrete (in Pounds)

Shank diameter	Penetration: 0.145" minimum	Type of loading	Concrete compressive strength (psi)	
			2000	3000
0.145"	1⅛"	Pullout	90	115
		Shear	160	225
0.177"	1⁷⁄₁₆"	Pullout	150	205
		Shear	250	285
0.205"	1¼"	Pullout	220	280
		Shear	390	445

NOTES:

Values based on low-velocity shot and stone aggregate.

Fasteners should not be located less than ½ inch from the edge of the concrete. Minimum spacing of fasteners should be 4 inches.

One-third increase for wind should not be included. These values do not take into account the capacity of the connected steel member, which might control. Capacity in shear should not exceed the following steel bearing capacities against the fastener shank:

	Shank Diameter		
Steel Thickness	0.145"	0.177"	0.205"
0.0478"	210	321	372
0.0359"	197	241	279

tening systems commonly are used to attach steel track to concrete slabs and foundations. The holding strength of the fastener in concrete depends on the compressive strength of the concrete, the shank diameter, depth of penetration, spacing, and edge conditions.

FIGURE 3.12 Use the powder-actuated stud/runner driver to drive fasteners into concrete.

Contact the manufacturer for more information on capacity and spacing requirements.

Headed or threaded drive pins are available with knurled shanks to increase their holding power in structural steel (Figure 3.12). The load might require the use of washers to prevent the fastener from prematurely pulling through the sheet steel.

Powder-actuated fastening tools use 0.22- to 0.38-caliber cartridges to provide the powder load. Correct fastener penetration is obtained by matching the required energy level to the job requirements. In low-velocity models, the force of the powder load acts on a central piston that drives the fastener. The velocity of the fastener can be changed to suit material densities by load selection and/or by placing the fasteners at different depths in the barrel bore. The proper powder load is based on the base material into which the fastener is being driven. The available energy levels are listed in manufacturers' catalogs for low-velocity models.

When a fastener is driven into concrete base materials, the holding power results primarily from a compression bond between the concrete and the shank of the fastener. Upon penetration, the fastener displaces the concrete which then tries to return to its original form and thus exerts a squeezing effect on the shank. The amount of compression increases in relation to the depth of penetration and the compressive strength of the concrete.

NOTE: Employ extreme care when operating either a pneumatically driven or powder-actuated tool. They can be dangerous if not operated as directed in the owner's manual.

CHOOSING BOLTS AND ANCHORS

Bolts and anchors commonly are used to fasten steel framing to masonry, concrete, and other steel components. Except for some proprietary anchors, it is not necessary to predrill

TABLE 3.7 Maximum Size of Bolt Holes
(in Inches)

Nominal bolt diameter, d	Standard hole size (in.)
Less than ½"	d + ½₂
Greater than ½"	d + ⅟₁₆

holes. Table 3.7 lists the maximum standard hole sizes for bolts in steel. Provide washers for oversized or slotted holes. Oversized or slotted holes in the direction of the anticipated load generally are not permitted unless evaluated by the designer. When the bolt or anchor is loaded in tension, washers might be required to prevent the anchor from prematurely pulling through the material.

Expansion Anchors. Expansion anchors in concrete or concrete block develop holding strength primarily from a compression bond between the concrete and the anchor shank. Drill a hole in the concrete that is the same size as the anchor. Install the anchor in the hole and expand it by turning the nut on the shank. A sleeve on the shank expands to provide the compression bond. Provided there is adequate embedment depth, the amount of compression increases with increased concrete compressive strength and the expanded anchor diameter. Expansion anchors are used in conjunction with hot-rolled steel plate, angles, or channel to supplement the bearing capacity of cold-formed steel framing.

Figure 3.13 shows a typical anchor bolt embedded in wet concrete.

Figure 3.14 shows a variety of steel-to-concrete single- and double-pour edge and corner installations. Install anchors with the bend-embedment line at the cold joint between the slab and the foundation.

Bolts. Steel studs and joists are not prepunched with fastener holes, whether threaded or unthreaded. As a result, nuts and bolts generally are not used with this system. The major exception is anchor bolts.

Nails. Spiral-shank nails, which have been case hardened and phosphate treated to etch their surface for increased holding power, are used predominantly to fasten plywood subfloors to the top flange of the steel joist (Figure 3.15). After the nails pass through the wood and penetrate the flange, the threads make a quarter turn to cut a partial thread into the pierced steel, which firmly connects the subfloor and joist. Nails also are employed to fasten materials to nailable studs and joists that consist of two rolled channel sections welded back to back, with nailing grooves for the attachment of collateral materials.

Adhesives. Adhesives are optional when self-drilling screws are used to attach plywood to steel joists. The use of adhesives is recommended, however, when hardened screw-shank nails are employed for this application. With either fastening method, the long-term bonding efficiency of construction-type adhesives contributes to stable, nonsqueaking floor assemblies and acts as an acoustical insulator. Adhesives also are used to attach drywall materials and paneling to steel studs or to laminate panels together. In these situations, adhesives serve to eliminate some, but not all, of the mechanical fasteners.

NOTE: Employ only approved adhesives and use as specified by the manufacturer. Consider temperature and moisture conditions to assure that adhesive strength is maintained.

Wire Ties. Wire tying is sometimes an acceptable method to attach metal lath and other metal parts to framing components. Never use wire tying to join framing components in load-bearing or curtain-wall applications.

Proprietary Fasteners. Several proprietary fasteners, which make installations easier and simpler, are available to the steel framer. Figure 3.16 shows a metal fastener that is used to fasten the studs to the floor and ceiling track.

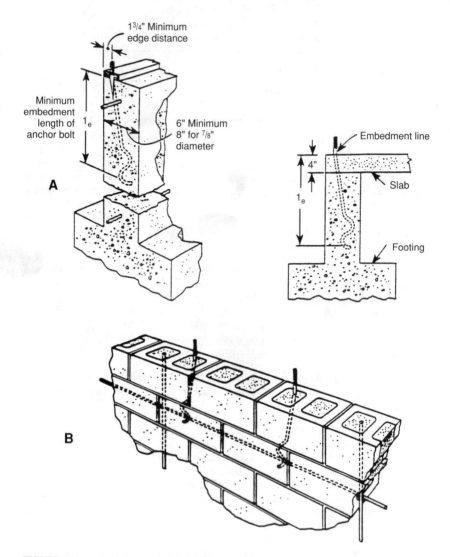

FIGURE 3.13 Embedded anchor bolt installed upon pouring concrete.

Another proprietary fastener for wallboard applications is the special edge-grip clip that mechanically attaches drywall panels over steel framing, steel furring on concrete block, or any other sound, level wall surface. Used primarily for predecorated drywall installations, the clip also can be used in standard gypsum-board installations.

Position the edge-grip clip on the back of the board and tap it into place with the installation tool (Figure 3.17). Drive the prongs into the board edges. Place the leading edge of each clip along one edge of the board and screw-attach it to the studs, furring, or other

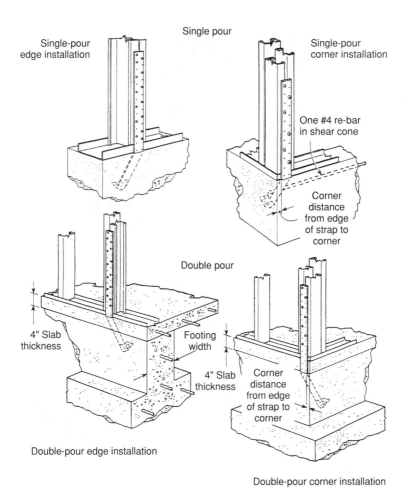

FIGURE 3.14 Typical steel-to-concrete single- and double-pour edge and corner installation.

FIGURE 3.15 Case-hard spiral-shank nails have good holding power.

wall surface. Slip the clips on succeeding panels under the previously applied panels. Fasten the top and bottom edges to the framing, furring, or wall (Figure 3.18). The edge-grip clips hold the board firmly in place. When using 4-foot-wide panels, use adhesive along the intermediate studs.

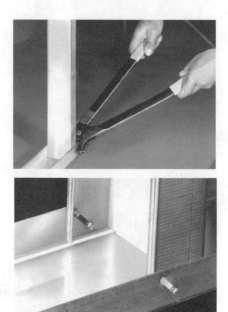

FIGURE 3.16   Steps required to install metal-lock
fasteners.

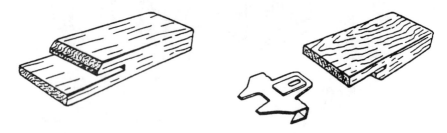

FIGURE 3.17   How a clip is held in an installation tool.

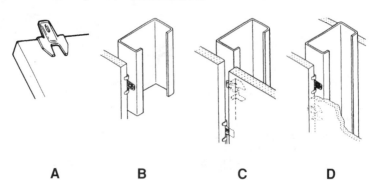

A   B   C   D

FIGURE 3.18   Installing an edge-grip clip.

WELDING TOGETHER MATERIALS

As defined by the American Welding Society (AWS), a weld is:

A localized coalescence (the growing together of the grain structure of the materials being welded) of metals or nonmetals produced by heating the materials to suitable temperature, with or without the application of pressure, and with or without the use of filler materials.

To elaborate on this definition, a weld is formed when separate pieces of material are fused together through the application of heat. The applied heat is hot enough to cause the pieces being joined to soften or melt. Once the pieces are heated, pressure might or might not be used to force them together. In some cases, filler material is added to the weld joint.

Welding can be used to prefabricate steel-framing components into panel assemblies and trusses, either at a contractor's shop or at the job site (Figure 3.19). Welding also is commonly used to attach site-fabricated or erected panels and to connect shelf angles to steel framing.

Arc-welding methods most often are employed to connect cold-formed steel shapes both in the shop and at the job site. Occasionally, contractors choose to connect cold-formed steel using resistance welding processes in the shop.

The two most common arc-welding methods are gas-metal arc welding (GMAW) and shield-metal arc welding (SMAW).

Using Gas-Metal Arc Welding

GMAW is an arc-welding process in which the arc is shielded by a chemically inert gas, such as argon or carbon dioxide (CO_2). This process often is called inert-gas welding for short or, when a consumable metal electrode is used, metal-inert-gas welding (MIG). In MIG welding, the electrode is mechanically fed into the arc (Figure 3.20). This type of welding is most economical in shop conditions. Good welds can be obtained with an AWS E-705-3, -5 or -6 wire electrode in a 0.030- or 0.035-inch diameter using CO_2, argon-oxygen, or argon CO_2 gas shielding.

For GMAW MIG welding, a wirefeed welder (Figure 3.21) with a capacity of 60 to 110 amps and approximately 23 volts variable output is suggested. This type of equipment usually requires 220-volt electric service. There are six basic welding techniques employed with MIG equipment (Figure 3.22):

Tack Weld. The tack weld is exactly that: a tack or relatively small, temporary MIG-spot weld that is used instead of a clamp or sheetmetal screw to hold the piece in place while making a permanent weld. Like the clamp or sheetmetal screw, a tack weld is always and only a temporary device. The length of the tack weld is determined by the thickness of the panel. Ordinarily, a length of 15 to 30 times the thickness of the panel is appropriate. It is important that temporary welds used to maintain proper panel alignment be done accurately.

Continuous Weld. In a continuous weld, an uninterrupted seam or bead is laid down in a slow, steady, ongoing movement. The gun must be supported securely so it does not wobble. Use the forward method: move the gun continuously at a constant speed and look frequently at the welding bead. Incline the gun between 10 and 15 degrees to obtain the optimum bead shape, welding line, and shield effect.

Maintain the proper tip-to-base metal distance and correct gun angle when making a continuous weld. If the weld is not progressing well, the problem might be that the wire length is too long. If this is the case, the metal is not penetrated adequately. For proper penetration and a better weld, bring the gun closer to the base metal. Handling the gun smoothly and evenly creates a bead of consistent height and width that has a uniform, closely spaced ripple.

A

B

FIGURE 3.19   Frames can be welded (A) in the prefab shop or (B) onsite.

Plug Weld.   To do a MIG plug weld, drill or punch a hole through the outside piece (or pieces) of metal, direct the arc through the hole to penetrate the inside piece, and then fill the hole with molten metal.

Spot Weld.   For a MIG spot weld, direct the arc to penetrate both pieces of metal, while triggering a timed impulse of the wirefeed welder.

Lap-Spot Weld.   The MIG lap-spot technique directs the arc so that it penetrates the bottom piece of metal. The puddle is allowed to flow into the edge of the top piece.

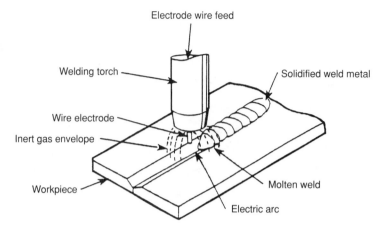

FIGURE 3.20 Basic MIG welding process.

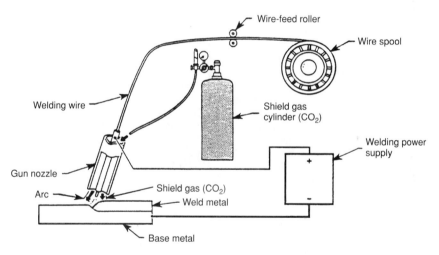

FIGURE 3.21 MIG welding equipment.

Stitch Weld. A stitch weld is a series of connecting or overlapping spot welds that create a continuous seam.

Using Shield-Metal Arc Welding

In the SMAW process, the arc supplies the heat needed to melt the base metal surfaces that are being joined and the filler metal in order to create a fuse. The filler metal is provided by a consumable, coated electrode that establishes the arc and gradually melts away and fills the joint. The shielding is accomplished by the melting of the coating on the electrode

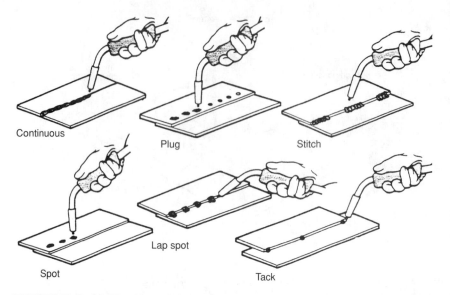

FIGURE 3.22 Basic MIG welding techniques.

(Figure 3.23). Good welds can be obtained with ³⁄₃₂- or ⅛-inch-diameter AWS-type E-6012, E-6013, or E-7014 rod. The heat setting varies depending on material thickness.

SMAW is a widely used welding process for cold-formed steel-framing systems because of its low cost, flexibility, portability, and versatility. Both the equipment and the electrodes are relatively inexpensive. The machine itself can be as simple as a 115-volt, stepdown transformer. The electrodes are made by a large number of manufacturers and are available in 1- to 50-pound packages (Figure 3.24).

The SMAW process is flexible in terms of the material thicknesses that can be welded and the number of positions that can be used. Materials from about 16 gauge to those that are several feet thick can be welded using the same machine with different settings. The flexibility of the process also allows materials in this thickness range to be welded in any position.

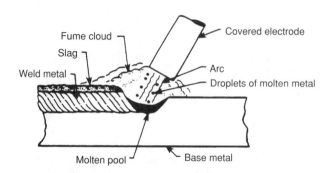

FIGURE 3.23 Shielded-metal arc welding.

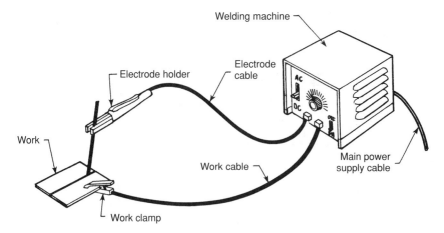

FIGURE 3.24   Shielded-metal arc welding equipment.

Employing General Welding Techniques

Because a hotter arc is recommended for welding galvanized metals, heat input is very important. Heat input is in direct relationship to the machine settings, travel speed, and how quickly the welded joints must be made to complete a repair. Because the hotter settings and slow travel speed work better on zinc-coated steels, use a skip welding technique. Weld short beads—no more than 1½ inches at a time—and allow the heat to dissipate before continuing to weld. Simply pause several minutes between welds.

The symbols for the popular fillet and groove welding methods (see Appendix C), include an arrow that connects the welding symbol reference line to one side of the joint. This side is considered to be the arrow side of the joint. The side opposite the arrow side of the joint is considered to be the other (far) side of the joint.

When a joint is illustrated on a drawing by a single line, and the arrow of a welding symbol is directed to that line, the arrow side of the joint is considered to be the near side of the joint.

For welds designated by the plug, slot, spot, seam, resistance, flash, upset, or projection welding symbols, the arrow connects the welding symbol reference line to the outer surface of one of the members of the joint at the centerline of the desired weld. The member to which the arrow points is considered to be the arrow-side member. The remaining member of the joint is considered to be the other side member.

Fillet Welds.   The dimensions of fillet welds are shown on the same side of the reference line as the weld symbol and are shown to the left of the symbol (Figure 3.25A). When both sides of a joint have the same size fillet welds, one or both can be dimensioned as shown in Figure 3.25B. When both sides of a joint have differently sized fillet welds, both are dimensioned (Figure 3.25C). When the dimensions of one or both welds differ from the dimensions given in the general notes, both welds are dimensioned. The size of a fillet weld with unequal legs is shown in parentheses to the left of the weld symbol (Figure 3.25D). The length of the fillet weld, when indicated on the welding symbol, is shown to the right of the weld symbol (Figure 3.25E).

Flare-Groove Welds.   Flare-groove welds extend only to the tangent points of the members (Figure 3.26). The size of a groove weld with a specified effective throat is indicated by showing the depth of the groove preparation. The effective throat appears in

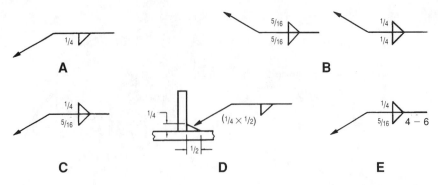

FIGURE 3.25 Typical fillet welds.

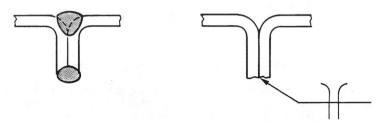

FIGURE 3.26 Typical flare-groove weld.

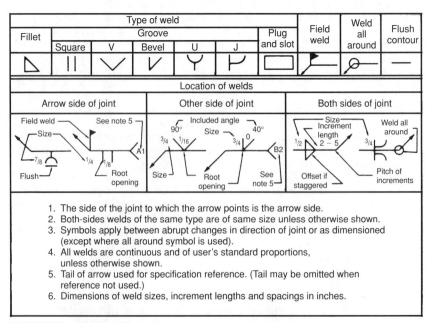

1. The side of the joint to which the arrow points is the arrow side.
2. Both-sides welds of the same type are of same size unless otherwise shown.
3. Symbols apply between abrupt changes in direction of joint or as dimensioned (except where all around symbol is used).
4. All welds are continuous and of user's standard proportions, unless otherwise shown.
5. Tail of arrow used for specification reference. (Tail may be omitted when reference not used.)
6. Dimensions of weld sizes, increment lengths and spacings in inches.

FIGURE 3.27 Standard welding symbols.

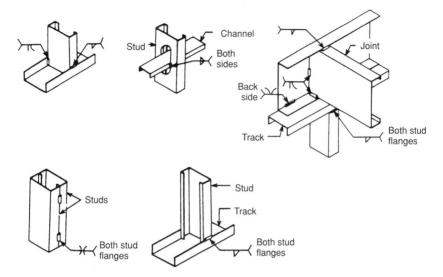

FIGURE 3.28 Welding symbols as they appear on a welding assembly drawing.

parentheses and is placed to the left of the weld symbol. The size of square-groove welds is indicated by showing the root penetration. The depth of chamfering and the root penetration is read in that order from left to right along the reference line.

Figure 3.27 gives the standard welding symbols used for steel framing. Figure 3.28 shows how the symbols appear on the working drawings to indicate where and how to weld.

CHAPTER 4
PRECONSTRUCTION PREPARATION

Advance planning by the framing contractor or steel framer can save time and material costs and result in a sounder and better-appearing job.

The general contractor or builder usually is in charge of the actual building of the structure. It is his or her job to follow the architect's or designer's plans and specifications, as well as the owner's desires. The general contractor subcontracts the work to a steel-framing contractor. Generally, it is the responsibility of the framing contractor to follow the plans and specifications as far as they concern the building of the structure. Failure to do so can result in callbacks and financial loss to the builder of the structure.

Many of the problems described in association with steel-framing installations are not the result of failures by the steel framer, framing contractor, builder, general contractor, or even the architect. For instance, structural movement can cause cracks in the drywall of the finished job. Structural movement and most cracking problems are caused by deflection under load, the physical alteration of materials due to temperature and humidity changes, seismic forces, or a combination of these factors. Descriptions of the major factors follow.

- Wind and seismic forces cause a cyclical shearing action on the building framework, which distorts the rectangular shape to an angled parallelogram. This distortion can result in the cracking and crushing of partitions adjacent to floors and structural ceilings.

- All materials expand and contract with temperature increases and decreases. In tall steel-framed buildings, thermal expansion and contraction might cause cracking if exterior beam components are exposed or partially exposed to exterior temperatures. Since interior columns and joists remain at a uniform temperature, they do not change in length.

- Many building materials expand when they absorb moisture from the surrounding air during periods of high humidity. They contract during periods of low humidity. Gypsum, wood, and paper products are more readily affected by hygrometric changes than are steel and reinforced concrete.

ESTIMATING MATERIALS AND PLACING ORDERS

Preconstruction preparation and planning involves the following:

- The plan or blueprint of the building
- Engineering planning
- Approval by local building officials and the general contractor or builder
- Blueprint cut-list

In the case of panelized and stick-built structures, a cut-list also must be made for the manufacturer.

Most steel-framing jobs begin with blueprints, which are prepared by the architect or design engineer (Figure 4.1). Whether you are a commercial or residential contractor, study the feasibility of each project you undertake. Decide if you want to take on the project.

Building with steel is not a radical departure from conventional homebuilding. Cold-formed steel offers a variety of construction options, from stick-by-stick framing approaches to preengineered and panelized systems. Steel also can be used in combination with more traditional building materials. Many builders using steel for the first time frame their interior nonload-bearing walls with steel, and then, when they are more familiar with the material, advance to constructing houses entirely framed in steel.

Regardless of the approach you use, you need to draw up a material list based on the engineered drawings. Be sure to closely review the details of the plans with a structural engineer to ensure structural integrity and the efficient use of materials.

With the preengineered system, the manufacturer prepares the steel shipping list. The list contains all the steel parts, the quantity of the pieces in the shipment, a description of each part, and usually part numbers, which might or might not be stamped on the parts. When the preengineered system parts arrive at the shop or job site, check to make sure the parts received match the manufacturer's shipping list.

With both stick and panelized construction systems, it is necessary for the framing contractor to prepare a *cut-list* to order the required steel materials. The data for the cut-list (Figure 4.2) usually can be found on the blueprint and includes information such as the manufacturer's part number, gauge, component, and the dimension of the piece. What gauge steel to use? The general rule of thumb for a single-story residence is that a 2 × 6 board beam can be replaced by a 6-inch, 20-gauge or a 2 × 4 lumber stud can be replaced by a 3-inch, 20-gauge. These substitutions might vary if extreme loading conditions are encountered in a two- or three-story structure (Figure 4.3).

Weight is another consideration. The weight of the truss has an effect on the design (Figure 4.4). To determine the weight each stud must carry, calculate the span. Let's use a 30-foot truss with a 15-foot span as an example. Each stud must support 15 square feet and hold up 2 feet of the truss (1 foot on either side) for a total of 30 square feet of roof (15 × 2). For this example, let's figure the weight per square foot at 35 pounds (10 dead weight, plus 25 live). Multiply this weight by the 30-square-foot requirement that each stud must support. The weight capacity of each stud must be 1050 pounds divided by 2 or 525 pounds.

To determine the number of square feet in the building, find the number of linear feet around the house on the foundation plan (Figure 4.5). If the framing is 16 inches on center, divide the linear feet by 16. If the framing components are to be spaced on 24-inch centers, divide the linear feet by 24. These figures give you a rough estimate of the number of studs, joists, rafters, and so on in the building.

This is a *rough* cut-list. A more exacting cut-list can be obtained by using the methods and forms detailed in Appendix D. While these procedures were formulated for preengi-

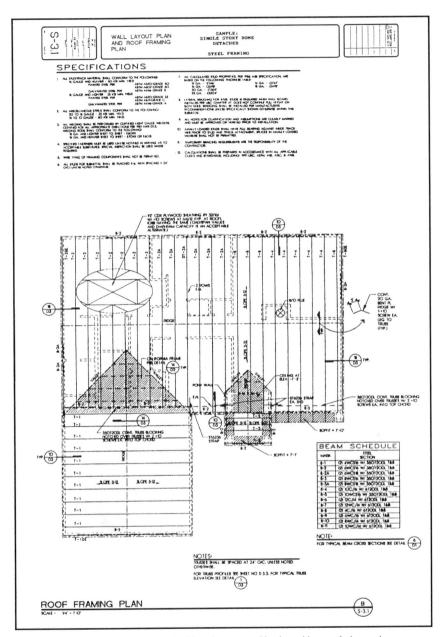

FIGURE 4.1 Framing jobs begin with the blueprints prepared by the architect or design engineer.

CUT LIST

K & S FRAMING
2283 CAMBRIDGE AVENUE
STEELTOWN, PENNSYLVANIA 19000

ITEM		MFG #	GAUGE	CUT SIZE	PART COMPONENT
1.	150	550XC	20ga	1'11 3/4"	Stem Wall
2.	20	5550WT	20ga	26'0"	" "
3.	50	350IC	20ga	1'3 1/4"	Pony Wall
4.	20	350WT	20ga	10'0"	" "
5.	40	2 x 2	18ga	1'11 1/2"	" "
6.	6	800WT	18ga	20'0"	Floor
7.	22	800EJ	18ga	24'11 1/2"	"
8.	40	078FC	20ga	20'0"	"
9.	150	600XC	20ga	8'0"	Exterior Walls
10.	30	600XC	20ga	7'4"	" "
11.	40	600WT	20ga	10'0"	" "
12.	20	600WT	20ga	19'0"	" "
13.	20	600WT	20ga	24'0"	" "
14.	20	600XC	18ga	8'0"	" "
15.	120	350IC	25ga	8'0"	Interior Walls
16.	20	350IC	25ga	12'0"	" "
17.	20	350IC	20ga	8'0"	" "
18.	60	350WT	25ga	10'0"	" "
19.	20	078FC	25ga	12'0"	" "

FIGURE 4.2 Typical cut-list prepared by the framing contractor.

FIGURE 4.3 Gauge of steel depends on the loading conditions at such location as the foundation columns.

FIGURE 4.4 The weight of a roof truss must be considered when determining stud size.

neered steel-framed residences, most of the information is helpful when making a well-documented estimate for both stick and panelized structures.

To get the necessary data for ordering, use the manufacturer's catalog. The wind-load table in the catalog generally is marked off in 12-, 16-, and 24-inch-on-center framing, with size ranges from ⅝ to 8 inches wide. As an example, let's assume that the walls of the proposed building are 8 feet tall and that the building code requirement states that a 15-pound wind load (about 80 to 85 miles an hour) is needed. To select the right materials, look under 20-gauge, which is the ideal gauge for most smaller residential homes, and check the capacity of the 24-inch-on-center framing. You find that it supports 1210 pounds. We need 1050 pounds for our example, so this stud size is well within the acceptable range. Complete instructions on how to use material found in a catalog is given in Appendix B. Symbols found in manufacturers' catalogs are explained in the List of Acronyms.

After finding the stud that meets the requirements, turn to the manufacturer's price list. Unfortunately most, if not all manufacturers, do not have a standard price list. Prices vary from region to region and from manufacturer to manufacturer. In addition, manufacturers give substantial discounts when buying in truckload quantities. A truckload of studs is about 40,000 pounds.

Once the framing cut-list is developed, take it to

- Steel-framing manufacturers
- Commercial building suppliers
- Home-improvement centers

Most suppliers can deliver standardized steel components that are specifically engineered for different framing needs. Steel ordered through a supplier can be cut to length and delivered to the job site, which can save considerable time. It also can be delivered in stock length and cut onsite to meet specific job needs.

Once the order is sent to the supplier, it is processed and the framing components are usually cut to size. The order is then packaged or bundled (Figure 4.6) and delivered to the

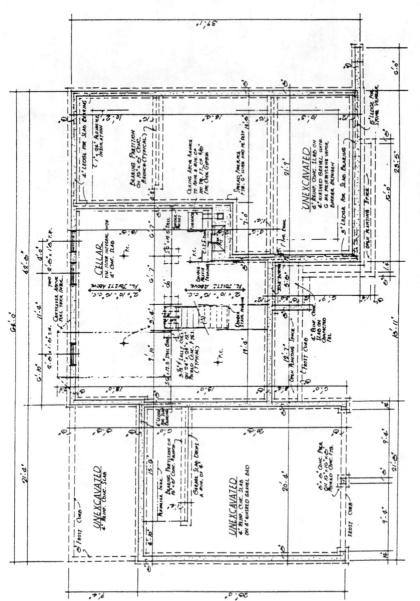

FIGURE 4.5 Typical foundation plan.

FIGURE 4.6 Bundling steel-framing components.

job site or the framing contractor's shop. Some larger shops are equipped with machinery (Figure 4.7) that can turn out studs, joists, track, or rafters.

Of course, before you send the cut-list to the manufacturer for processing, you should get a stamp of approval on the drawing and other items required by the local building code. Some codes require the approval of the following items in order to receive a building permit.

Shop Drawings.   Drawings should illustrate materials; shop coatings; steel thicknesses; fabrication details; details of the frame attachment methods; size, location, and spacing of frame fasteners; details of structural attachment methods; accessories and their installation methods; and critical installation procedures. Drawings might include plans, elevations, sections, and details.

Samples.   Samples should be representative pieces of all framing component parts and accessories. Unless otherwise specified, sample pieces should be 12 inches long and tagged with the name of the part and the manufacturer's name.

Calculations.   Submit engineering calculations or data that verify the framing assembly's ability to meet or exceed design requirements as required by local codes and authorities.

Certifications.   Include statements from the manufacturer attesting to the fact that the materials conform to the appropriate requirement as required by local codes and authorities.

FIGURE 4.7 Many large shops have machinery like this that turn
out steel components like studs, track, and joists.

Literature.   Submit all technical literature prepared by the framing manufacturer.
    There are steel-framing estimating concerns around the country that provide the fol-
lowing services:

- Engineering calculations
- Thermal-energy calculations
- Cut-lists of sheathing materials
- Customized installation instructions

- Conversions from architectural wood-framed plans
- Installation work-hour estimate forms to bid framing
- Cut-lists of steel components for walls, floors, and roofs
- Inspection checklist with recommended construction tolerances
- Specified steel products available through local material suppliers
- Material purchase orders that maximize precut items to minimize setup charges
- Accurate counts of screw and pin connectors with a printed purchase order
- Engineered details completed in three dimensions and plotted on D-size drawings
- Steel purchase orders with manufacturer's product designations

To use one of these services, fill out an order form similar to the one shown in Figure 4.8.

CALCULATING LABOR COSTS

Two components must be considered in order to properly estimate the labor costs on a steel-framing installation:

- Number of work hours required to perform the installation work
- Taxes and insurance costs associated with the total number of work hours allocated to the job

Let's consider how to estimate the work hours required to construct the framework for a steel-framed home. Figure floors, load- and nonload-bearing walls, roof, and interior work separately and then add the amounts together.

At this point, the definition of the term *work hours* must be clearly understood. For example, if a truss installation takes 2 workers ½ hour each to set a truss, the total number of work hours is 1. If 1 worker takes 4 hours to install bridging, the total number of work hours is 4. If it takes 4 workers 4 hours to unload and store materials, the total number of work hours is 16.

Once the total number of work hours is calculated, multiply that figure by the per-hour labor rate to get an estimate of the labor costs to complete the basic installation for that segment of the job.

Then add the *labor burden* to the labor costs in order to calculate total labor costs. On average, the labor burden increases labor costs by 25 to 30 percent. The labor burden is the percentage of payroll dollars the framing contractor is compelled to pay in taxes and the amount of premiums paid to insurance companies. The following taxes and insurances are examples of the components that make up the labor burden.

Unemployment Tax. All states levy an unemployment insurance tax on employers. The tax is based on the total payroll for each calendar quarter. The actual amount assessed varies according to the history of unemployment claims filed by employees of the company. This tax might range from 1 to 4 percent of total payroll dollars.

FUTA. This is the federal government's unemployment tax. This tax has averaged 0.8 percent of payroll dollars per employee up to a maximum dollar amount per year. The tax is periodically set by law.

Social Security and Medicare (FICA). The federal government also requires FICA payments averaging 7½ percent of payroll dollars per employee up to a maximum dollar amount per year as prescribed by law.

RESIDENTIAL PROJECT ORDER FORM

Contact _____   Phone _____
Company _____   Fax _____
Address _____
        (street)   (city)   (state)   (zip code)

1) Building Code: ☐ CABO, ☐ ICBO, ☐ BOCA, ☐ SBCCI, ☐ Other _____

2) 1st Floor Deck: ☐ Slab on Grade   ☐ With 1st. Floor Joist Framing System

3) Floor Areas: (all dimensions are taken from outside to outside of walls; SF indicates square foot)
   Garage: ☐ SF   1st Floor : ☐ SF   2nd Floor: ☐ SF

4) Wall Stud & Floor Joist Member Spacing: (check one for maximum spacing)
   ☐ 16"o.c.   ☐ 19.2"o.c.   ☐ 24"o.c.   Other: _____

5) Deflection Criteria: (check one for minimum allowable deflection)
1st Floor:	☐ L / 480	☐ Architect	☐ Per Code	Other: _____
2nd Floor:	☐ L / 360	☐ Architect	☐ Per Code	Other: _____
Roof:	☐ L / 240	☐ Architect	☐ Per Code	Other: _____

6) Floor Joist Depth: (lap joists are spliced over bearing points, full spans are not spliced)
   1st Floor: ☐ Per plan   ☐ Specify 8", 9 1/4", 10", 12",14"   ☐ Lap ☐ Full Span
   2nd Floor: ☐ Per Plan   ☐ Specify 8", 9 1/4", 10", 12",14"   ☐ Lap ☐ Full Span

7) Wall Stud Depth: (check for stud depths, heights & finishes)
   Exterior: ☐ 2 1/2" ☐ 3 1/2" ☐ 5 1/2" ☐ 8" ☐ As Per Plans
   Interior: ☐ 2 1/2" ☐ 3 1/2" ☐ 5 1/2" ☐ 8" ☐ As Per Plans
   Sheathing: ☐ OSB ☐ Plywood ☐ Foam ☐ Ext. Gypsum   Other: _____
   Wall Heights: ☐ Per Plan   1st Floor: _____ ,   2nd Floor _____

8) Roof Framing Configuration: (check for style and spacing of roof system)
   Roof Type: ☐ Truss   ☐ Stick or Hip Framed
   Spacing: ☐ 16"o.c. ☐ 19.2"o.c. ☐ 24"o.c.   Other: _____

9) Window & Door Schedule: (a window and door schedule must accompany each order with the following information)

Example for R.O.'s:	TYPE	SIZE	ROUGH WIDTH	ROUGH HEIGHT	HEADER HEIGHT
	1	30 x 310	3'-1 1/8"	3'-11 1/2"	6'-10 1/2"

ALL DIMENSIONS ARE TO BE STEEL TO STEEL (add 1-1/2" to each side if using wood inserts)

10) Framing Manufacturer: (if a specific material manufacturer is not listed, one will be chosen by TSN, Inc.)
   Name of Manufacturer: _____

Price Calculation: Total Area in Square Footage (See Item 3 Above)

Package # I - Material Take-Off   ☐ SF x 0.19 $ ☐

Package # II - Engineering and Material Take-Off   ☐ SF x 0.59 $ ☐

FIGURE 4.8 Typical estimating service form.

Worker's Compensation Insurance All states require that employers carry this insurance to cover employees in the event of a job-related injury. The cost of this insurance is a percentage of payroll dollars based on the occupation of the employee. Secretarial jobs have a low rate when compared to high-risk jobs, such as those found at construction sites. Construction jobs carry a rate that is usually between 5 and 10 percent of payroll dollars.

Liability Insurance This insurance protects the steel-framing contractor in the event of an accident. Most general contractors require that a certificate of insurance be presented by subcontractors prior to beginning work. The cost of this insurance is based on total payroll and is dependent on the location of the company, the work performed, history of claims by the company, and the required liability limits.

These examples show why the labor burden must be included in every labor cost estimate. Each state and local government passes its own laws and sets its own percentages for these taxes and individual companies regulate their own insurance rates. These costs must be updated frequently to avoid underpriced jobs, reduced profit margins, and even losing money.

A computer, plus the proper software, is a handy tool for creating steel-framing cutlists. It also can be used to estimate labor and material costs. Software is available to make estimates for residences, offices, apartment buildings, shopping malls, and other commercial structures.

USING PANELIZED AND NONPANELIZED CONSTRUCTION METHODS

In a stick-for-stick approach, the spacing of studs ranges from 16- to 24-inch centers depending on the drywall requirements. Panelized and modular systems are generally engineered to require fewer studs.

A list of important things to do that you should keep in mind when installing stick-built members follows.

• Align track accurately on the supporting structure and fasten it to the structure as shown on the shop drawings.

• Evenly butt track intersections.

• Plumb, align, and securely attach studs to the flanges or webs of the upper and lower tracks. Squarely seat axial-loaded studs in both the top and bottom tracks.

For panelized structures, it is important to keep in mind the following requirements.

• Design panels to resist construction and handling loads, as well as live loads.

• Avoid causing permanent distortion in any member or collateral material when handling and lifting prefabricated panels.

• Make all stud-to-track connections before hoisting the panel.

• When it is necessary to splice the track between stud spacings, place a piece of stud in the track and fasten it with two screws or welds per flange to each piece of track.

• Ensure complete bearing under the tracks to provide for load transfer in axial-loaded assemblies.

• Attach the panel to the structure as shown on the shop drawings.

• Align all panels to provide continuity of any wall/floor surface.

To fabricate cold-formed structural framing systems economically and successfully, a fabricator should have an adequately equipped shop. Fabrication of the panel in a fabricating shop (Figure 4.9) often offers many advantages over site fabrication.

• Site conditions might not allow for the fabrication of panels and the storage of materials due to space limitations. Some jobs require that the panels be erected right from the panel transporting truck.

FIGURE 4.9 A fabricating shop has advantages over onsite fabrication.

- Shop fabrication of panels can commence while site work and footings are being poured. The panels then can be erected as soon as the foundation or slabs are ready to receive them.
- Shop fabrication can continue through bad weather.
- Fabrication costs often are lower in the shop. Workers are more efficient in a shop environment, where an assembly-line system can be established.
- The work quality in a shop is often better than in the field. Shop welding in a controlled atmosphere is superior to field welding, especially when the gas metal arc welding (GMAW) process is used.

Figure 4.10 shows a typical shop layout with equipment and supplies conveniently located for the efficient movement of materials. Here are some general and basic recommendations for designing an efficient shop.

- Establish a proper in-process flow. When planning your assembly and fabrication shop, arrange for materials to move smoothly and rapidly from one station to the next.
- Avoid delays or bottlenecks by placing tools, framing materials, and supplies close to the worktables and within easy reach of the workers.
- Timing is important. The assembly and fabrication should proceed with the needs of the job. For example, balance the fabrication to meet the speed of erection.

In general, the following space, equipment, and tools are needed to set up a fabrication shop:

- Minimum of 2400 square feet of floor space, plus a yard to store materials and panels
- Raw storage area approximately 12×32 feet, plus one horizontal rack
- Panel storage area that is a minimum of 12×40 feet
- Jig fixtures that are 9 feet 6 inches wide and 32 feet long for residential construction and 12 feet wide and 40 feet long for commercial construction

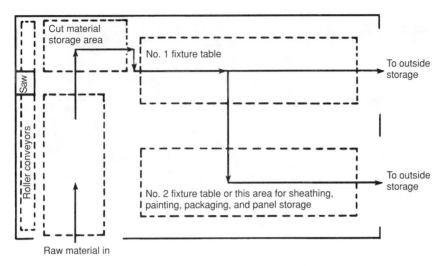

FIGURE 4.10 Typical fabrication-shop layout.

- Four fitting clamps or air/hydraulic cylinders on each table to properly seat and hold the track section over the studs
- Two metal-inert-gas (MIG) welders per fixture table that have .030 to .035 wire, rectifiers, wirefeed guns and spot-time attachments
- An overhead track on which to mount the wirefeed and spot timer
- One 3200 rpm saw fitted with a 16-inch metal cutting blade
- One forklift with a boom to move the panels
- 30-foot roller conveyor in 10-foot sections
- Vise grips
- Canvas shields for weld flash protection
- Paint touchup equipment
- Miscellaneous tools as required

The fixture tables are key to successful assemblies and fabrications. Most shops have a truss table for assembling trusses (Figure 4.11A), wall panels (Figure 4.11B), prefabricated trusses (Figure 4.11C), and a subcomponent assembly table (Figure 4.11D). To minimize the investment, the tables can be built in the shop from cold-formed steel-framing materials. Figures 4.12 and 4.13 show typical fixture tables that are extendable from 7 feet 6 inches in width and 32 feet in length. These tables should meet the requirements of most residential and commercial construction work.

ARRANGING FOR HANDLING AND TRANSPORTATION

Steel framing normally can be handled and transported without expensive equipment. A forklift and flatbed truck usually are sufficient.

FIGURE 4.11 The following fixture tables are important to the success of fabrication shop: (A) truss assembly table; (B) wall panel bench; (C) prefabricated truss table; and (D) subcomponent assembly table.

When a job requires a full load of material, the delivering carrier, whether a common carrier, a steel hauler, or the framing manufacturer, will probably use a flatbed truck or trailer. The material comes in bundles or on pallets. Pallets generally are approximately 2 feet wide, 2 feet high, and the length of the material (from 8 to a maximum of 40 feet). Bundle weight generally does not exceed 4000 pounds. A forklift can handle most unloading jobs quite easily (Figure 4.14). A spread beam with end slings also can be used with a hoist or crane (Figure 4.15).

C

D

FIGURE 4.11 *(Continued)*

As the materials are unloaded, place them conveniently around the site. For one-story structures on a concrete slab, place the material approximately where it will be used. If the building is two or more stories high, stack the material for the upper floors around the building where it can be used later, floor by floor. For curtain-wall applications, hoist the material to its appropriate floor with a job site crane or take it up the freight or construction elevator. Small dollies or wagons might be required to get the material from the elevator to the appropriate area.

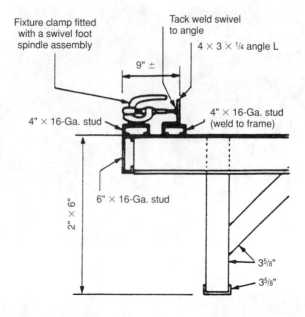

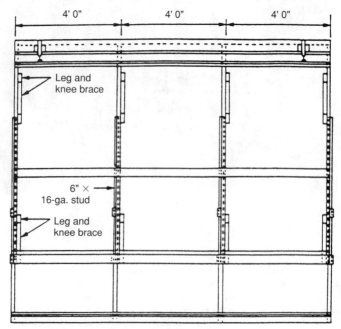

FIGURE 4.12 Typical shop-built fixture table.

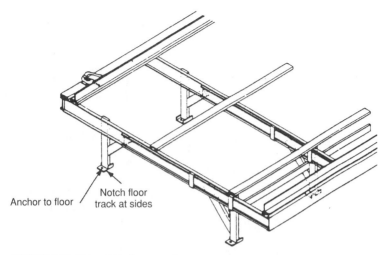

Anchor to floor / Notch floor track at sides

FIGURE 4.13 Fixture-table floor mounting.

FIGURE 4.14 A forklift is valuable for moving all types of materials.

If panelized construction is used on a slab or floor, tilt the panels into place by hand. If the panels are oversize, or must be attached to outside supports, a portable crane might be needed. Panels for low-rise buildings can be hoisted into place by a forklift with extendable forks, or lifted by a small boom or cherrypicker.

FIGURE 4.15   Using spread-beams and slings with hoists or cranes.

Material going to the shop can either be cut to length or fabricated. Assuming the material is to be fabricated into panels, unload it in the fabrication area. Move the materials from the storage area to the saw with a forklift or overhead crane. The feed into a saw generally is from a conveyor, roller-type stands, or table. The cut lengths continue from the saw on roller stands and can be stockpiled on material wagons by hand (Figure 4.16).

Then move the wagons to the assembly area where the panels are fabricated. Again, handling might be by hand. The completed panels then can be moved to a staging area or loaded directly onto a truck or trailer. Depending on the panel size and weight, use either manpower, a forklift, or an overhead crane for loading. Once the material is unloaded at the job site, further transporting rarely is required.

Shop-fabricated material must be moved from the shop to the job site. The distance might vary from a few miles to many. When deciding which transporting equipment to use, consider the following seven factors.

Panel Finishes.   If the panel is free of collateral materials, such as insulation, metal lath, plaster, and so on, it can be transported flat. If not, vertical shipment is required. Unfinished or frame-only panels obviously are not as delicate as those with a finishing material on one side. These panels can be handled and transported without extra precautions and more panels can be loaded and moved at one time. A semifinished panel requires more care when loading, blocking, tying down, unloading, and erecting it. Difficult site conditions might justify the extra expense of this type of panel.

Transporting Distance.   The ideal distance is a short haul from 0 to 50 miles. Few problems should be encountered in this distance. A medium haul from 50 to 100 miles can be practical if the project is large and existing shop space is available and economical. A long haul of 100 or more miles usually is not practical, particularly if the loads are over-sized or the job-to-shop route is through a congested urban area. A long haul through a rural or suburban area, on the other hand, might be justified.

FIGURE 4.16 Push handcarts are handy in a shop.

Panel Size. Panels that are 8 feet or less in width usually can fit in a standard truck bed. If you haul panels that are more than 8 feet wide, you might need a special permit.

Permits. Check with local authorities regarding the following categories. The usual requirement is provided under each category.

- Overwidth
  ~up to 8 feet no problem
  ~8 to 10 feet permit only
  ~10 to 12 feet permit, plus lead car
  ~12 to 14 feet permit, plus lead and tail car when freight rate is approximately 200 percent of normal
- Overlength
  ~up to 4 feet permissible with red flag
  ~over 4 feet permit, plus additional tail lights
- Overheight
  ~more than 12 feet might require permits
  ~careful route selection to avoid most underpasses
- Restrictions
  ~often are limited from dawn to dusk
  ~no weekends
  ~no holidays

Panel Weight. Weight usually is not prohibitive since the panels are steel-framed. The unfinished weight is approximately 2 pounds per square foot. The volumetric space usually is filled before weight limitations are reached.

Number of Trips. If many trips or oversized loads are required through a congested area, the freight expense might be prohibitive and job site fabrication should be considered.

FIGURE 4.17   A flatbed trailer can be used for short-haul jobs.

Job Site Conditions   If the site has ample space, the panels could be delivered and stockpiled. If there is restricted space, the panels might have to be erected off the truck as they are delivered.

Once an evaluation is made to satisfy drayage considerations, and depending on available equipment, the delivery vehicle might be one of the following.

- A flatbed trailer, pulled by a pickup truck, is recommended for small panels, short-distance hauls, and relatively small jobs (Figure 4.17).
- Consider using a regular flatbed truck with a 16- to 20-foot bed for small jobs that are going a medium-haul distance with average or possibly oversized panels.
- For a large project that requires several medium- to long-distance hauls, with possibly oversized panels, consider using a tractor and semitrailer.
- A lowboy tractor-trailer generally is recommended for semifinished panels, oversized panels, or oversized panels that are going to be transported vertically.

The contractor's first consideration is whether to fabricate onsite or in the shop. Factors influencing this decision include the actual physical size of the project, the amount of steel framing required, the size of the shop, and the availability of equipment.

RECEIVING STEEL-FRAMING MATERIAL

Make provisions for storage before accepting delivery of the material. Once the material arrives at the job site or shop, it generally is unloaded from the truck with a large forklift or crane. When a crane is used, only one signal person should direct the crane operator. Both the crane operator and the signal person should have a thorough understanding of the signals used.

During the unloading process, the signal person should remain in clear view of the operator and maintain a clear view of the load. The signal person should alert everyone within the cone of activity before signaling the operator to hoist or handle a load. The signal person also should tell other workers when the handling operations have ceased and work can proceed with *normal precautions*.

It is the signal person's responsibility to make sure the load is not hoisted above anyone. When unloading overhead workloads, keep the following safety precautions in mind.

- Frequently inspect chokers, slings, spreaders, and all other equipment that is used to handle overhead tools. Inspect for cuts, kinks, excessive wear, or anything that might result in failure. Never use this equipment as a ground for welding.
- Properly balance all loads and use taglines to maintain balance and control (Figure 4.18).
- Tightly choke bundles to prevent loose pieces from shifting or sliding free.
- Handle loads at minimum heights to safely clear obstructions.
- Check overhead loads before hoisting to be certain that they are free of loose bolts, nuts, washers, rocks, mud, or anything that might fall free during or after handling.

Hoisting equipment should have load-limit capacities prominently noted on the equipment for the operator to see. Give careful attention at all times so as not to exceed the stated limits.

When unloading the framing material with either a forklift or crane, the following guidelines assure a smooth operation.

- Have a definite unloading and storage plan that is understood by the entire crew. Least-handled material is usually the safest-handled material.
- Make everyone aware of the fact that loads might shift in transit and again when unloaded.

FIGURE 4.18 When unloading, make sure loads are properly balanced.

- Make it a practice for everyone to keep hands and feet clear and out from under the material.
- Use blocking when placing and removing chokers and slings.

Assemble the frame that will be raised last first. Stack the assembled frames on top of one another in order, last frame to be raised on the bottom with the first frame to be raised on the top. Place wooden blocking between each of the assembled frames to make it easier to handle the frames.

NOTE: When stacking light-gauge material, do not set unlike materials on top of each other. Scatter the bundles so that none of the unlike frames are stacked on top of each other and that each frame is easily accessible when needed.

Depending on the width of the frame, a spreader bar might be required to lift and place the frames conveniently when using a crane (Figure 4.19). A 50-foot tagline tied to each end of the frame acts as a guideline and helps control the frame while it is in motion.

The maximum weight of any bundle should not exceed 4000 pounds. Band each bundle with steel bands (Figure 4.20). All shipments normally are loaded with a 5000-pound capacity, all-terrain forklift, that is equipped with pallet forks.

Check each loaded truck for completeness. After the truck is unloaded, check each piece against the shipping documents and note any shortage or damage. Report the discrepancies to the manufacturer within 72 hours.

Most flatbed shipments allow four hours for off-loading, or *demurrage*. Thereafter, penalties are assessed at an hourly rate by the trucking company directly to the consignee. Most shipments can be offloaded with a forklift in less than two hours.

When offloading, place bundles on wooden blocks. Elevate one end to allow standing water to drain. Cover loosely to protect from rain and to allow air circulation to evaporate any condensation. Place all bundles far enough apart so that personnel can walk between them and get material as needed.

FIGURE 4.19 Use of a spreader bar in lifting operations, such as this truss.

FIGURE 4.20 Typical bundle of steel-framing components.

FOLLOWING CONSTRUCTION-SITE SAFETY PROCEDURES

The most important single factor in avoiding the pain, suffering, and expense inherent in construction accidents is the development, by each worker, of a *safe worker* attitude. All job site personnel should follow good safety practices.

While the information contained herein might serve a useful purpose by reminding you of various factors that you should consider to maintain safe job site operations, it does not begin to cover all aspects of an adequate safety program. The requirements of applicable federal, state, and local statutes, ordinances, rules, and regulations have not been considered. Compliance with these regulations is mandatory. No attempt has been made to be all inclusive. Do not assume that all precautionary measures, particularly those adopted due to unusual job site conditions or to ensure safe use of a particular piece of equipment, have been adequately addressed in this book.

Employing Emergency Accident Procedures

Establish and review safety procedures with workers at the job site. Post the phone numbers of doctors, ambulance services, hospitals, fire departments, and police at the job site.

Have emergency equipment, such as first aid kits, available at the job site. Have fire extinguishers available in the vicinity of heat-producing operations. Fire extinguishers should be appropriate for the types of fire they might need to extinguish. Regularly recharge and inspect the extinguishers. During cold weather, protect the extinguishers with antifreeze additives. Instruct all workers in the operation of the extinguishers and how to select the proper extinguisher for each type of fire.

In general, *Class A* fires refer to those involving wood, textiles, or rubbish. A fire extinguisher using a chemical solution, water-antifreeze mixture, a wetting agent, foam, a loaded stream, or ABC dry chemicals works to combat such fires (Figure 4.21).

FIGURE 4.21 ABC dry chemical fire extinguisher.

Class B fires are those that involve ignited grease, motor vehicles, or flammable liquids. Fire extinguishers that use foam, carbon dioxide, dry chemicals, or ABC chemicals are effective with these fires.

Class C fires involve live electrical equipment. Only extinguishers that use carbon dioxide, dry chemicals, or ABC chemicals are effective with Class C fires.

Taking Personnel Safety Precautions

Workers should wear gloves, a hard hat, appropriate protective gear, and ankle-supporting soft-sole shoes. When conditions warrant, workers should wear a dust mask. Until the crew is comfortable handling cold-formed steel, it is recommended that workers wear a pair of lightweight gloves to protect their hands from small cuts and abrasions.

In order to lessen the possibility of injury from lifting, follow these rules.

- Place feet in a position close to the load that provides maximum balance.
- Keep the back straight and as near to vertical as practical while bending the knees and reaching for the load.
- Keep elbows as straight as possible so that when the load is grasped the body is over the load and the back is as erect as possible (Figure 4.22).
- Lift using the leg muscles. Do not lift by bending the back.
- Grasp the load firmly and take care to keep the load from slipping.
- Once the load is lifted, do not unnecessarily readjust or shift grips to prevent the load from slipping free.
- When changing the direction of travel, move the whole body, including both feet. A sudden twist of the upper body can cause back injuries.

Workers should be quick of mind, able to think on the job, and capable of looking for safety hazards without constant direction. Discourage horseplay. The use of alcohol and drugs should not be tolerated under any circumstances. Workers with distorted perception and/or hangovers are a dangerous risk and should not work.

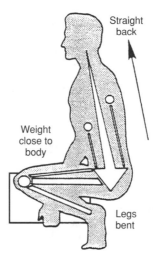

FIGURE 4.22 Use leg muscles to lift any size load; never use the back.

Holding Safety Meetings

Hold a job site meeting at least once a week to review that week's work and any violations of good safety practices that occurred during the previous week. Appoint an individual from the work crew to check for adherence to safe practices. Rotate the responsibility among the crew members. Bring in safety engineers from insurance companies to review your program.

Considering Job Site Safety

The safety rules on a steel-framed job site generally are the same as those for traditional construction. Some of the important job site safety considerations follow.

Weather Conditions. Follow changing weather conditions so that work on unprotected portions of the house is not attempted during unsuitable weather conditions. Personnel should not climb or work on scaffolds during snow or rainstorms, or in high winds.

Sanitary Facilities. Have sanitary facilities available at the job site.

Drinking Facilities. Provide adequate drinking water and salt tablets.

Temporary Material Storage. Prepare a temporary storage area before materials are delivered. Maintain the area to protect the building material from adverse weather conditions and abusive handling. Proper stacking is required for safety, easy access, and to reduce material damage.

Barricades and Warning Signs. Install and maintain properly prepared warning signs and barricades in all dangerous areas.

State and Local Safety Requirements. Since all state and local building code requirements must be followed, check them in advance.

Housekeeping Keep the job site clear of trash and excess material that is not being used. Cleaning up on a regular basis is the best practice. Potential buyers are impressed with an orderly job site.

Secure the Job Site. Take advance precautions to make sure the job site is well protected, both during the day and at night.

Storage and Use of Hazardous Materials. Handle dangerous and hazardous materials with extreme caution. Store such materials according to the manufacturer's written recommendations.

Removal of Materials. Maintain an organized work area with access in and about the work site at all times. Periodically remove trash. On large jobs, designate specific areas for trash storage. Each trade should remove its trash no later than the completion of its part of the job.

Temporary Electrical Service. Centrally locate the temporary power pole as close to the work area as possible. Typically, the temporary hookup of the power pole is furnished by the builder or is the responsibility of the electrical subcontractor. Make sure that you determine this when negotiating your subcontract.

Bracing. Brace the structure at all times and before raising the next part. Secure the structure with temporary or permanent bracing before removing the raising equipment and at the end of each day, on weekends, or during any other shutdown periods.

Bolts and Thread Chasers. Make up all joints and place all bolts in place before removing any supports. If the bolt threads are such that it is necessary to use a thread chaser, then replace the bolts with new ones.

Roof and Floor Openings. Caution all workers regarding roof and floor openings.

Climbing Structures. Do not allow workers to slide down columns. Use ladders to get on and off the structure.

Unauthorized Personnel. Do not allow persons not connected with the job on the premises. Provide all authorized persons with hard hats, appropriate protective gear, and a constant escort. You are liable for any injury that might occur to the people you allow on the job site.

CHAPTER 5
STEEL FLOOR FRAMING

Generally, the work of the framing contractor and his or her crew begins when the foundation is ready to accept the steel framing, which usually is about seven days after it is poured. When the framing crew arrives at the job site, the foundation should be completed and ready for the steel framing to be erected. All the necessary raw materials, the framing tools and equipment, and a set of working drawings should be properly stored and available at the job site.

The four phases of steel framing include the floor assembly, exterior and interior load-bearing walls, nonload-bearing interior partition walls, and the roof rafters and supports. The structural steel skeleton consists of floor framing, wall framing, and roof framing (Figure 5.1). These structural elements, or phases, together form a strong framework that serves two basic purposes:

• Withstand the effects of wind, water, and temperature

• Provide a base to which enclosing exterior materials and attractive interior materials can be attached

Joists are the main horizontal structural members in floor framing. The joists lay on the foundation wall, columns, or footings (Figure 5.2). To increase the strength of the steel frames, joists are covered with a sheet material such as plywood, particleboard, or oriented strand board (OSB). The sheet material applied across the floor joists is called a subfloor. Underlayment board is usually laid on top of the dry subflooring. The finished floor material, such as carpet, vinyl, and tiles, are then attached to the underlayment board.

LAYING SOUND FOUNDATIONS

A sound foundation is essential to the construction of a structure, whether it be a single-family dwelling or a skyscraper. Certain basic principles apply to the construction of footings and foundations (Figure 5.3). The foundation provides a stable base for the building.

FIGURE 5.1 Steel skeleton showing floor, wall, and roof assemblies.

FIGURE 5.2 Floor joists in place on foundation.

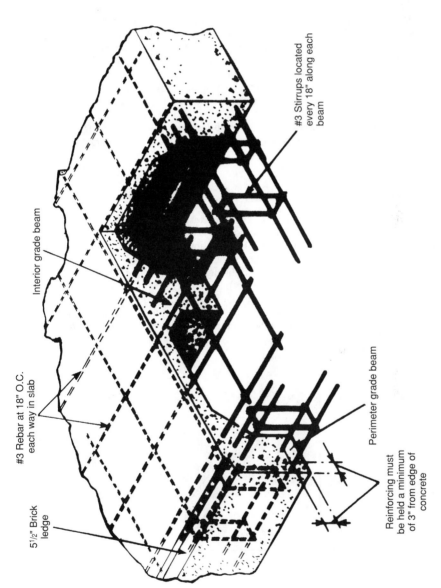

#3 Stirrups located every 18" along each beam

Interior grade beam

#3 Rebar at 18" O.C. each way in slab

5½" Brick ledge

Perimeter grade beam

Reinforcing must be held a minimum of 3" from edge of concrete

FIGURE 5.3 Typical foundation construction.

It distributes the weight of the building and applied loads over the bed area to prevent the building from settling unequally. When soil conditions are poor and do not provide good bearing, footings are spread over a greater area. The foundation must be laid out accurately and must be of proper design to support the load above.

The design of the foundation is the responsibility of the architect, but the framing contractor is responsible for laying out and constructing the formwork. The exception to this is when the framing contractor is acting as both the general contractor and framing contractor, such as when building a preengineered home.

Although the setting of forms and the pouring of the foundation usually is left to a masonry crew, the framing contractor and framer are responsible for verifying the dimensions of the foundation. Remember that steel framing can be used on any type of foundation: slab-on-grade, crawl-space, or basement.

Since the foundation dimensions are so important to the structural integrity of the building and the framing, the contractor or framer must check the foundation for the following items.

Squaring the Foundation

All corners must be perfect right angles. If leveling instruments are not used, the layout can be made or checked using the 3-4-5 method, which is based on the pythagorean theorem. According to this famous mathematical proposition, a triangle with sides of 3, 4, and 5 units, or multiples of these figures, form a right triangle. If one side of a right triangle measures 3 feet, and the side meeting it at a right angle measures 4 feet, then the diagonal, or *hypotenuse* side, measures 5 feet (Figure 5.4). For greater accuracy, however, multiples of these numbers are used, such as 6, 8, and 10 feet or 9, 12, and 15 feet.

When checking for squareness, it is a good practice to verify the diagonal measurements of every major jog, or *offset*, in the foundation. To do this, pick the diagonal you

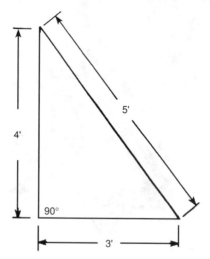

FIGURE 5.4 Pythagorean theorem is useful in foundation layout work. If the base is 3 feet and the altitude 4 feet, the hypotenuse is 5 feet. Multiples of these distances can be substituted.

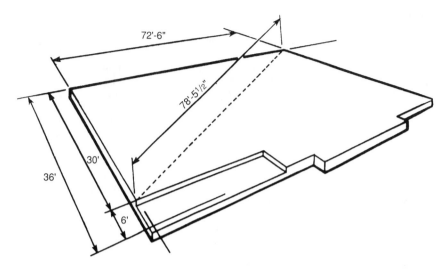

FIGURE 5.5 Checking the squareness of a foundation.

wish to verify. For this example, let's choose B1 to G10. Determine the length and width dimensions from B1 to G10 (Figure 5.5). The given length is 72 feet 6 inches. Then calculate the width: 36 feet − 6 feet = 30 feet. There are two ways to calculate the diagonal.

Formula. (Length)^2 + (Width)^2 = (Diagonal)^2 (Length × Length) + (Width × Width) = Diagonal × Diagonal or 72 feet 6 inches × 72 feet 6 inches + 30 feet × 30 feet = 5256.25 plus 900 = 6156.25. Now, using a calculator, take the square root of 6156.25, which is 78.46 or 78 feet 5½ inches.

Rise-and-Run. If you are familiar with rise-and-run method, then you know that 30 feet is the rise and 72 feet 6 inches is the run. The diagonal is 78 feet 5³⁵⁄₆₄ inches, which is rounded to 78 feet 5½ inches. This technique is quite fast with a construction calculator.

Leveling the Foundation

Using a surveyor's transit, check the overall levelness of the foundation. Shim the column bases if the foundation is more than ⅛ inch low at the anchor bolts. Chisel or grind high spots at the anchor bolts.

Avoiding Thread Damage

Use a wire brush to remove any concrete from the anchor-bolt threads and use a thread chaser to correct any thread damage.

Spacing the Anchor Bolts

Using a 100-foot steel tape, measure several bays at a time to determine if the anchor bolts have been accurately placed. Start at the endwall anchor bolts.

Grading Elevations

Grade elevations usually are related to a reference point, called the *benchmark*, which is assigned an arbitrary elevation of +100 feet. Other elevations around the job site are determined by comparing them to this reference point. Higher points have elevations greater than +100 feet; lower points less than +100 feet. The benchmark must be stable and stationary. One corner of the building often is made the benchmark against which grade elevations are checked.

If there is a squaring problem, adjust the starting line to best suit the dimensions of the foundation. A ⅛-inch change to this baseline can cause as much as a 6-inch change on a long house.

Snap a chalkline from sidewall to sidewall at the anchor-bolt centerline. Hold the chalkline on each end and then have one worker in the center hold and snap the line. The anchor-bolt centerline usually is 3 inches in from the endwall stud-wall line. Locate the anchor-bolt line at the opposite endwall, or wherever the section of wall ends. Snap the line again. If you use a slab or a wall that is more than 8 inches wide, locate the inside of the steel line on both sidewalls and snap a line. This line usually is 8 inches in from the outside steel line.

Measure and mark the centerline bay spacing on the line. Using a framing square, carefully locate the bolts per the drawing. Mark a + at each location. With the framing square, make a line 1¾ inches left or right of the + and then square it to the inside steel line. This allows you to position the steel accurately on 8-foot centers and at the correct dimension to the opposite sidewall. Install the anchor bolt or chemical anchors per the manufacturer's instructions.

DETERMINING JOIST ELEVATIONS

Floor joists carry structural loads and therefore require a minimal amount of engineering. In residential construction, joists often are supported only at each end and do not have any concentrated loads imposed on them. Use Table 5.1 as a guide to determine which steel joists can be substituted for wooden joists. A qualified engineer must give special consideration to the bearing points, web crippling, connections, etc. Figure 5.6 illustrates a typical floor-framing plan.

The joist elevation is determined by the concrete wall form. The elevation is important for under-joist headroom and for stair construction. A finished-floor-to-finished-floor measurement is a key measurement for stair design (Figure 5.7). The frames and restraining forms for all openings that appear later in the foundation, such as basement windows

TABLE 5.1   Maximum Allowable Clear Span C Joists Subjected to 40 psf Live Load and 10 psf Dead Load (In Feet)

Joist spacing	6" 18 ga.	6" 16 ga.	8" 18 ga.	8" 16 ga.	9¼" 16 ga.	9¼" 14 ga.	10" 16 ga.	10" 14 ga.
12" O.C.	12.4	13.3	15.6	16.8	18.9	20.4	20.2	21.8
16" O.C.	11.3	12.1	14.2	15.3	17.2	18.5	18.4	19.8
24" O.C.	9.8	10.6	12.2	13.3	15.0	16.2	16.0	17.3

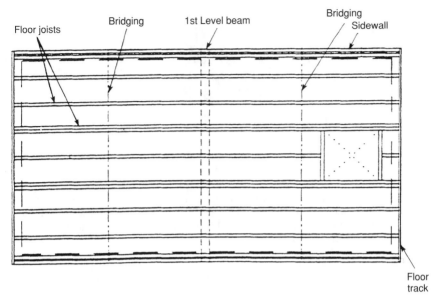

FIGURE 5.6 Plan view of floor joists.

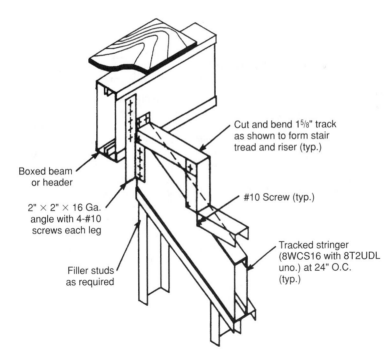

FIGURE 5.7 Basement stairwell details.

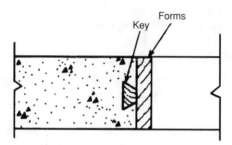

FIGURE 5.8   The form-key-concrete relationship for keying a door into the foundation wall.

and pipe and door openings, must be placed within the concrete wall forms prior to pouring the concrete. Such openings appear as lined openings when the forms are removed. Basement window openings usually are fitted with frames that are keyed to the concrete and made ready to receive window sashes.

Pipe openings for vents or services are designed to provide access that is secure against water and air movements. Door openings usually include a form designed to be keyed to the concrete wall (Figure 5.8) and braced to withstand the pressures of fluid concrete. The key remains in the concrete when the forms are removed.

The elevation of a door opening in a basement wall is set in relation to the expected finished outside grade and the inside landing. Flights of stairs can lead from the landing to the basement and to a first-floor level. The elevation of the landing provides a uniform unit of rise for the stairs.

ANCHORING THE FOUNDATION

To prevent movement from wind pressure, firmly attach the superstructure to the foundation with anchor bolts (see Chapter 3). The pressures created by the wind can cause translation, overturning, and rotation. *Translation* is the action by which a building is moved in a lateral direction by strong winds. *Overturning* is the uplifting of light superstructures by wind forces. To avoid overturning, use both anchor bolts and framing anchors. Attach the anchors to the individual framing members, usually on 4-foot centers.

Rotation occurs when a light, framed building pivots on its foundation. Rotation sometimes is caused by nonsymmetrical high winds. To prevent rotation, firmly anchor the steel track to the foundation with anchors that possess adequate shearing strength.

Attach the runner track to the foundation wall with ½-inch anchor bolts spaced on 6-foot centers (Figure 5.9). Place the bolts so that their threaded ends extend above the top of the foundation. If the foundation wall is constructed of concrete blocks, place the anchor bolts in a core filled with mortar. To keep the mortar from falling out of the core, place a metal lath below the hollow core.

In typical residential floor systems, fastener attachments usually are located in the built-up beam assembly and from the:

- End stiffener to rim joist or track
- Connection clip to joist
- Joist to web stiffener

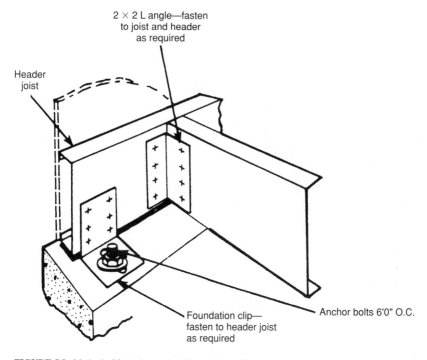

2 × 2 L angle—fasten
to joist and header
as required

Header
joist

Foundation clip—
fasten to header joist
as required

Anchor bolts 6'0" O.C.

FIGURE 5.9 Method of fastening track to foundation wall.

- Bridging to joist
- Solid blocking to joist
- Joist hanger to joist
- Joist to overlapping joist

LAYING OUT THE JOISTS

After the foundation work is complete, it is time to frame the structure. The floor frame, which is assembled on the foundation, consists of track, beams, joists, columns, and sub-flooring. Basically, the floor framing is the same whether the structure is built on a slab or over a full basement. Proper construction and sound materials are important to ensure that the house is built rigidly. Figure 5.10 shows a floor frame under construction.

Installing the Perimeter Track

The runner, or track, rests on or is attached to the foundation walls and forms the bearing surface for joists and studs. Provide a uniform and level joist-bearing surface at the foundation walls by using shims and/or nonsetting grout or caulking compound. Cut the track

FIGURE 5.10   Steel-framed floor assembly under construction.

to length. Joist members usually can be ordered cut to length for most applications. Field-cut shorter track members (Figure 5.11).

As shown in Figure 5.12, the track can be securely anchored to the continuous sidewall of the foundation, to the base wall if on a concrete slab, flush with the top of the foundation, or joist connected. Use one of the following attachment methods.

- Stub nails or powder-driven anchors on concrete and masonry (Figure 5.13)
- Metal concrete inserts using ⅜-inch type-S-12 panhead screws
- Mushroomhead concrete sleeve anchors
- Similar concrete fasteners (see Chapter 3)

For a wooden fountain, use 1¼-inch type-S ovalhead screws or 12d nails.

Check the layout of the level with a chalkline, plumb bob, level, and/or laser alignment tool to make sure the layout matches the engineer's or architect's plan. When making any measurement, make sure the tape is pulled tight.

Before starting to install the perimeter track, insert a strip of building paper or felt between the track and the concrete. This helps isolate the foundation wall from the metal if there is a moisture problem. To help hold the building paper in place, apply two beads of a silicone or acrylic caulking compound around the outside perimeter. This caulk, when depressed with a form board, fills any irregularities in the concrete.

After the caulk and paper are in place, install the track from the middle of its run or in the middle of the piece. Never start at one end and move to the center. The end might be out of alignment. Secure the track or runners with fasteners spaced on maximum 24-inch centers and secure 2 inches from each end. Be certain there is no gap in the track since it could cause a floor squeak later.

FIGURE 5.11 Shorter track members field-cut with a chopsaw.

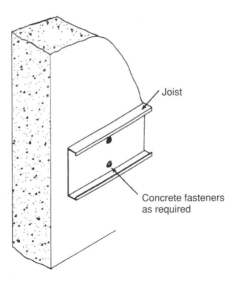

Joist

Concrete fasteners
as required

FIGURE 5.12 Floor joist parallel to a continuous
wall.

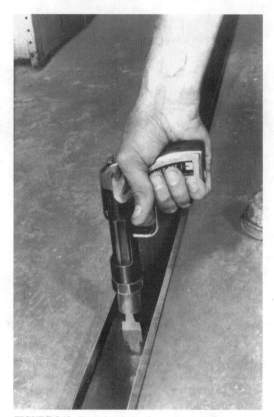

FIGURE 5.13 Fastening the track runner.

NOTE: In crawl-space foundations (Figure 5.14), stubs, and other crawl-space basements, it might be necessary to cut the track and pull the track leg up to allow for ventilation of the cavity. Hold the insulation out of the vent. For long, continuous runs of track, it might be necessary to splice the runner as shown in Figure 5.15.

Selecting Floor Joists

Floor joists are the main supporting members that carry floor loads to the track sills and beams (Figure 5.16). They must be of sufficient strength and stiffness to withstand the intended load with a minimum amount of deflection. Joists that lack the proper stiffness can cause nails to pop in the walls and ceilings. Floors tend to give when occupants walk across them, which causes an uneasy feeling. Joists are commonly spaced on 16-inch centers, but other spacings can be used. The size and spacing must conform to local building codes.

Figure 5.17 shows a joist connected to a built-up joist. Provide additional joists or *blocking* adjacent to exterior and interior walls, openings, and elsewhere when needed to support ceiling construction as indicated by the plans and specifications. Provide an additional joist under parallel partitions when the partition length exceeds one-half the joist span and for all floor and roof openings that interrupt one or more span members, unless

FIGURE 5.14 Crawl-space or pony-wall foundation under construction.

otherwise noted. Fix the ends of the joist to prevent rotation and buckling. Steel joists also can be attached to other steel joists with steel joist hangers and/or clip angles when allowed by manufacturer's specifications. The ends of the joists can be set in pockets that are formed in masonry block or concrete. Adequately reinforce, level, and block these joists.

Reinforce the ends of steel joists to adequately stiffen the joist web and transfer loads to the supports as provided by the plans and specifications. At the points of stress, reinforce the web with a length of stud, track, or angle section. The web stiffener, as shown in Figure 5.18, can be field cut and then welded or screwed into place.

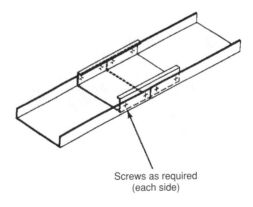

Screws as required
(each side)

A

FIGURE 5.15 Splicing runner track.

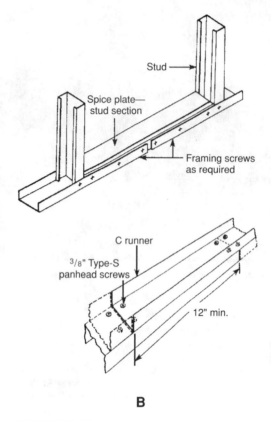

B

FIGURE 5.15 *(Continued)*

PLACING BEAMS AND COLUMNS

When the joists must cross a long span, use beams to support them. The joists are load-bearing members and must be supported firmly as shown by the upset deep beam and upset shallow beam in Figure 5.19. Rest the joist ends on the ends or sides of the foundation walls (Figure 5.20). At intermediate points, support them with columns or studs (Figure 5.21).

Steel beams usually have a cross section that resembles the letter I. Formerly called I beams, they are now called S beams. Wide-flange beams are known as W beams. The size of the beam is determined by the load it must carry. The cross section and length are specified in the construction plans and on the cut-list.

Figure 5.22 illustrates how a joist can be fastened over a continuous beam and span. A continuous beam is a structural system that carries load over several spans using a series of rigidly connected members that resist bending moment and shear. The loading can either be concentrated or distributed along the lengths of members. The underlying structural system for most bridges is a set of continuous beams.

Many steel-framing contractors prefer to use interior-lapped joist joints over the beam (Figure 5.23). Connections between the lapped joists are not moment resistant. Meeting the minimum lap dimension requires an increase in the web-crippling capac-

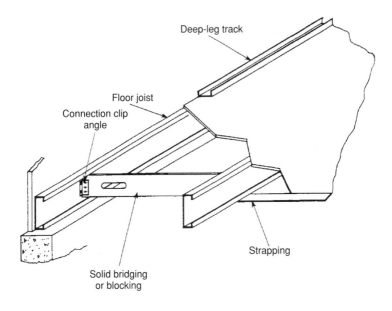

Deep-leg track

Floor joist
Connection clip
angle

Strapping

Solid bridging
or blocking

A

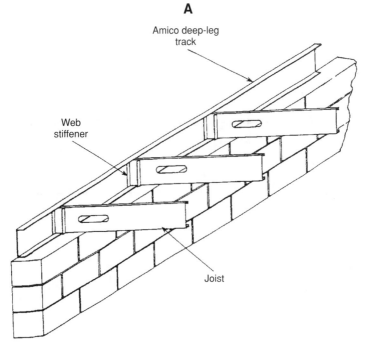

Amico deep-leg
track

Web
stiffener

Joist

B

FIGURE 5.16 (A) Floor joist bearing on solid bridging and (B) floor joist bearing on masonry foundation.

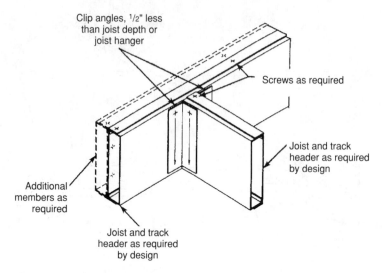

FIGURE 5.17 Typical joist header to a built-up joist.

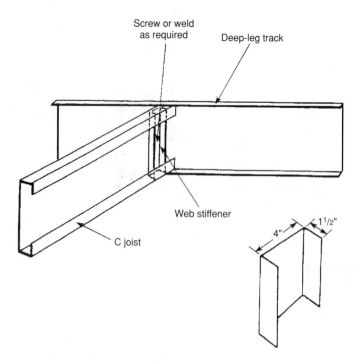

FIGURE 5.18 Installed web stiffener.

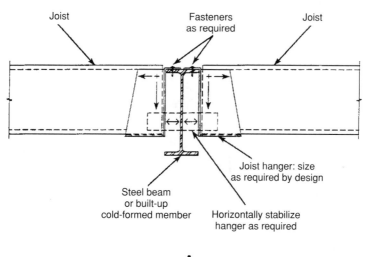

A

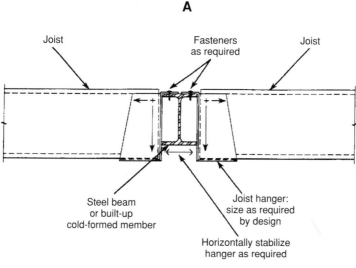

B

FIGURE 5.19 Joist supported by (A) steel upset deep beam , (B) upset shallow beam.

ity as the reaction is not applied to the ends of the joists. This situation might require a web stiffener.

Joist hangers can tie the beam to the steel-framing network. Use a web stiffener and clip angle to stabilize the bottom of the joist in this setup (Figure 5.24). The single- and double-joist arrangements shown in Figure 5.25 are located parallel to the exterior wall. A nonload-bearing interior wall also is generally set parallel to the joists (Figure 5.26).

Certain floor construction designs, for example, have sunken floor areas (Figure 5.27) and various cantilevered floor effects (Figure 5.28). Bar-joist bearings between studs can be

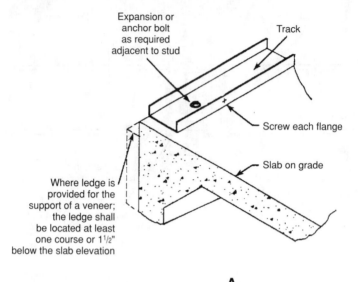

Expansion or
anchor bolt
as required
adjacent to stud

Track

Screw each flange

Slab on grade

Where ledge is
provided for the
support of a veneer;
the ledge shall
be located at least
one course or 1½"
below the slab elevation

A

Joist

Concrete fasteners
as required

B

FIGURE 5.20 Track support at the (A) base of the wall, (B) flush with the top of foundation, (C) another plumb with top of foundation, (D, E, F) support methods when the track is centered below the wall.

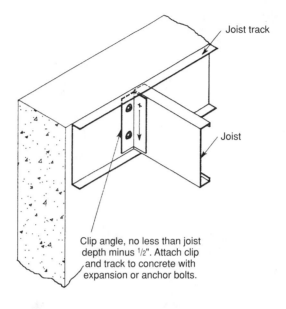

Joist track

Joist

Clip angle, no less than joist depth minus ¹/₂". Attach clip and track to concrete with expansion or anchor bolts.

C

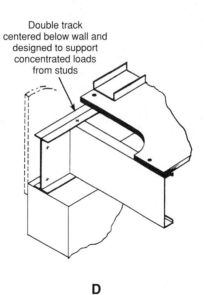

Double track centered below wall and designed to support concentrated loads from studs

D

FIGURE 5.20 (Continued)

Track and joists
centered below wall
designed to support concentrated
loads from studs

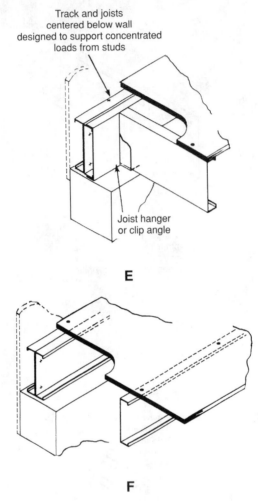

Joist hanger
or clip angle

E

F

FIGURE 5.20 *(Continued)*

installed as shown in Figure 5.29. Web stiffeners might be required at load-bearing points and/or at reaction points. Fasten joined members together on maximum 12-inch centers.

Using Steel-Beam and Column-Load/Span Tables

Guidelines for selecting and designing beam and column bearings can be found in *Residential Steel Beam and Column Load/Span Tables (RG-936)*, which is published by the American Iron and Steel Institute (AISI). The publication contains data on residential steel beams and columns to support floor-system spans or tributary widths that range from 6 to 24 feet in 2-foot increments.

The tables include floor-system dead loads of 10, 15, and 20 pounds per square foot (psf) to account for various finishes or superimposed loads. The weight of the steel beams

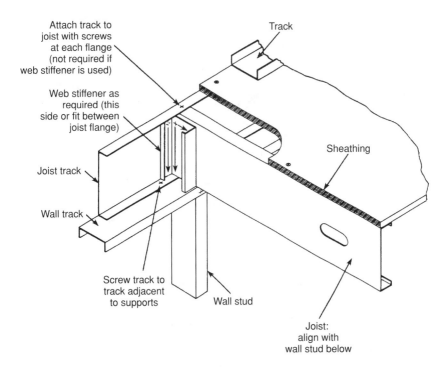

Attach track to
joist with screws
at each flange
(not required if
web stiffener is used)

Track

Web stiffener as
required (this
side or fit between
joist flange)

Sheathing

Joist track

Wall track

Screw track to
track adjacent
to supports

Wall stud

Joist:
align with
wall stud below

A

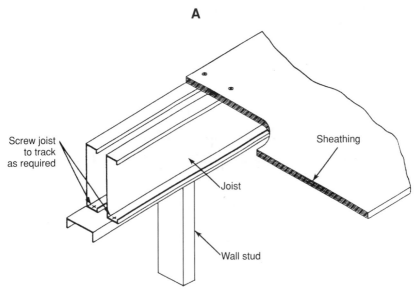

Screw joist
to track
as required

Sheathing

Joist

Wall stud

B

FIGURE 5.21 Two methods of intermediate support for joists.

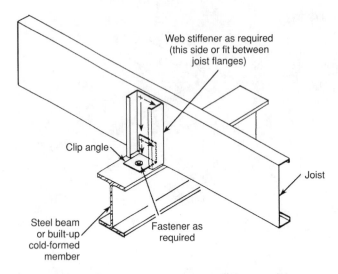

FIGURE 5.22 Method of fastening a joist over a continuous beam.

is included in the calculations. In addition, the weight of the interior bearing wall that supports the second and third floors (as applicable) also is included. Roof-system loads are not included. It is assumed that the roof system spans between the exterior bearing walls. If the roof system is supported by the interior bearing walls and the roof loads do not exceed the floor loads used in the tables, the tables can be used assuming the roof is considered an additional supported floor.

A typical example of dead-load weight is given as follows:

- Floor dead load (psf)
  ~wooden flooring = 2.5

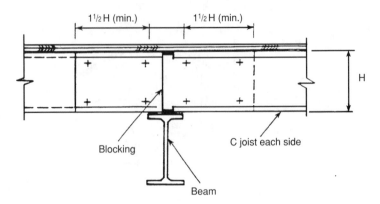

FIGURE 5.23 Interior lapped-joist over a beam.

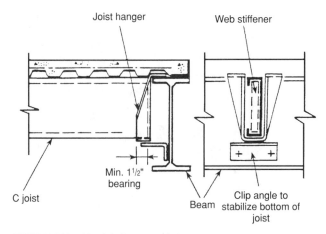

FIGURE 5.24 Using joist hangers to tie beam.

~subflooring = 2.0
~10-inch steel joists 24-inch × 1 foot, 5 inches = 1.5
~½-inch gypsum wallboard ceiling = 2.0
~miscellaneous and mechanicals = 2.0
~Total = 10.0 psf
• Interior-bearing wall dead load (plf)
~3⅝-inch steel studs 24-inch × 8-feet 5 inches = 5.0
~top and bottom track = 2.0
~½-inch gypsum wallboard each side × 8 feet = 32.0
~Total = 39.0 plf

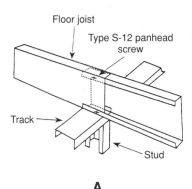

A

FIGURE 5.25 (A) Single- and (B) double-joist setups.

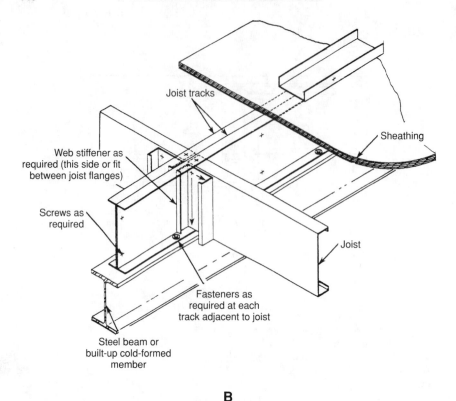

B

FIGURE 5.25 *(Continued)*

Tabulating Beam-Span Values

Beam designs are based on single, simple-span conditions (Figure 5.30A). This approach is conservative for multiple, continuous spans. These spans generally are controlled by allowable moment capacities, but in some cases are limited so as not to exceed a live-load deflection limit of L/360. In the case of continuous beams, span values limited by live-load deflection are somewhat conservative.

The tabulated beam-span values are based on a minimum of 1¾ inches of bearing along the axis of the beam at each beam end and 3½ inches at each column when beams are continuous over the column. In some cases, beam spans have been limited in order to maintain these bearing requirements. A bearing width perpendicular to the beam axis, in addition to the beam-flange width, also might be required when the beam is supported on materials other than steel.

Using Column Tables

Do not use the column tables for column designs that change in section profile between the top plate and base plate, such as screwjack or adjustable-height columns (Figure 5.30B). These columns generally are proprietary and require testing to establish load-carrying capabilities. Contact the manufacturer for the design capacities of these columns.

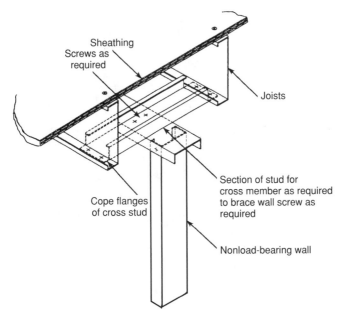

Sheathing
Screws as
required

Joists

Cope flanges
of cross stud

Section of stud for
cross member as required
to brace wall screw as
required

Nonload-bearing wall

FIGURE 5.26 Nonload-bearing interior wall.

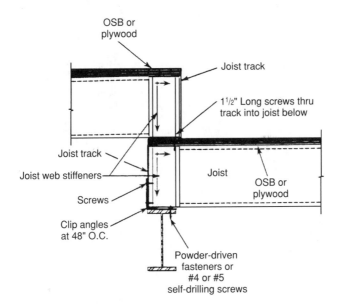

OSB or
plywood

Joist track

1¹/₂" Long screws thru
track into joist below

Joist track

Joist web stiffeners

Joist

Screws

OSB or
plywood

Clip angles
at 48" O.C.

Powder-driven
fasteners or
#4 or #5
self-drilling screws

FIGURE 5.27 Details of the floor joists at a sunken floor.

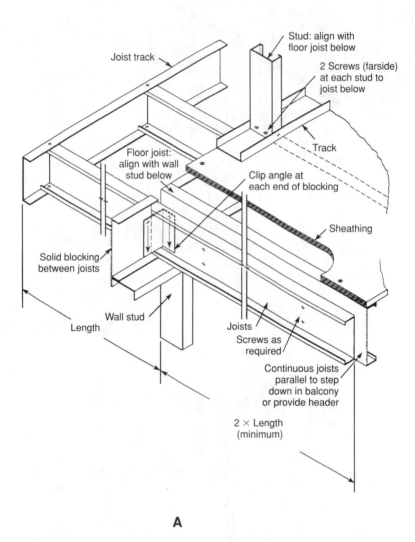

A

FIGURE 5.28 Various cantilevered floor details: (A) balcony with stepdown; (B) cantilevered floor at wooden deck balcony, (C) cantilevered floor joist at flush balcony floor, and (D) cantilevered floor at stepdown balcony floor.

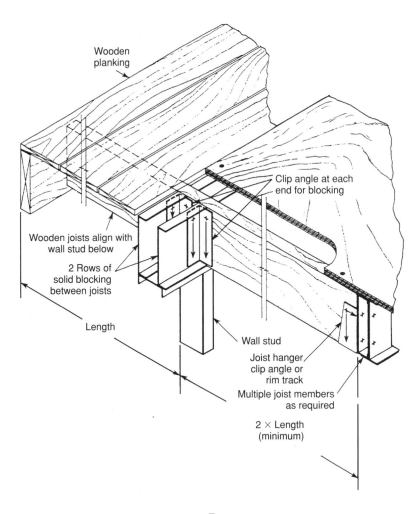

Wooden
planking

Clip angle at each
end for blocking

Wooden joists align with
wall stud below

2 Rows of
solid blocking
between joists

Length

Wall stud

Joist hanger,
clip angle or
rim track

Multiple joist members
as required

2 × Length
(minimum)

B

FIGURE 5.28   *(Continued)*

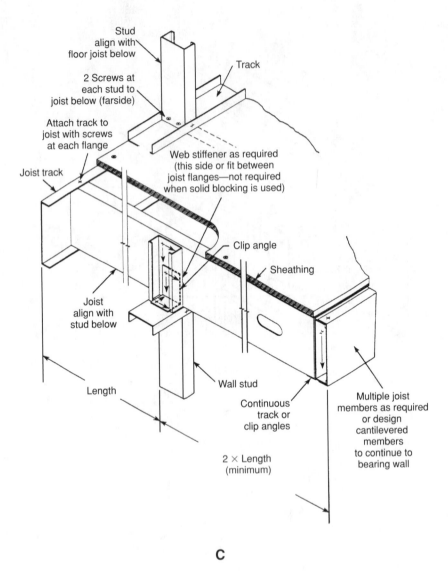

C

FIGURE 5.28 *(Continued)*

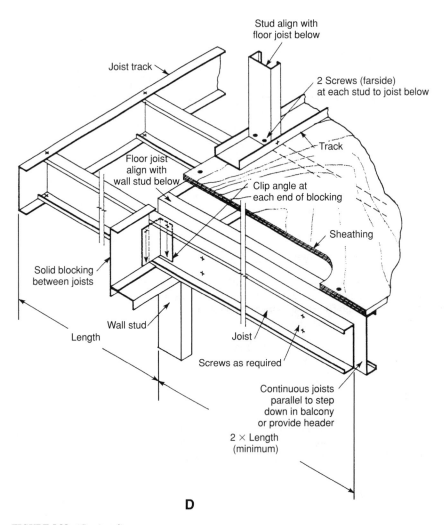

D

FIGURE 5.28 *(Continued)*

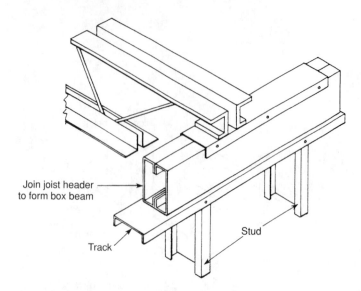

Join joist header
to form box beam

Stud

Track

FIGURE 5.29   Bar joist bearing between studs.

The column spacing dimensions listed in the column tables have been calculated assuming that the eccentricity of the total or resultant load is 1 inch or less from the column centerline. This eccentricity is provided to account for some moment induced into the column through minor eccentricities in beam bearing. Eccentricities might exist as a result of connection detailing, unequal beam spans on each side of the column, and unequal, or *pattern*, loads on each side of the column.

Copies of the steel beam and column tables can be obtained from AISI, 1101 17th Street, NW, Washington, DC 20036-4700. There is a slight charge for this publication.

FRAMING FLOOR OPENINGS

Cutting through joists to make openings for stairwells, chimneys, and fireplaces, considerably weakens the floor frame. To compensate for this, frame the shortened joists, called *tail joists*, against doubled headers. Tail joists, in turn, are framed against joists or doubled trimmer joists (Figure 5.31).

As already mentioned, most manufacturers provide holes in the webs of joist components to accommodate electrical and plumbing lines. Should larger holes be required for drainlines, vents, or other mechanical installations, reinforcement is required around the installed piping to protect it should the web collapse (Figure 5.32). As a rule of thumb, if more than 25 percent of the web is removed, reinforcement is required to add back the required strength.

For larger openings, building codes and regulations determine the required width and headroom of stairs and the clearances that must be placed around chimneys and fireplaces. The openings for these components must be laid out carefully and exactly as specified in the building plans. Errors are quite expensive to correct.

Residential Steel Beam Load/Span Tables - Wide Flange Beams

DL (psf)+	10
LL (psf)*	40

MAXIMUM SPAN FOR CENTER BEAMS SUPPORTING ONE FLOOR ONLY (no roof or attic loads) - L

TRIBUTARY WIDTH SUPPORTED BY THE CENTER BEAM - (A+B)/2

BEAM PROPERTIES (Min. F_y = 36ksi)

SIZE	I	S	Mc	Lc	RE	RI	6'-0	8'-0	10'-0	12'-0	14'-0	16'-0	18'-0	20'-0	22'-0	24'-0
W6x9	16.4	5.6	11.0	4.2	11	26	14.3 LL	13.0 LL	12.0 LL	11.3 LL	10.8 LL	10.3 LL	9.8 LL	9.3 LL	8.9 Mc	8.5 Mc
W6x12	22.1	7.3	14.5	4.2	18	36	15.8 LL	14.3 LL	13.3 LL	12.5 LL	11.9 LL	11.4 LL	10.9 LL	10.6 LL	10.2 Mc	9.8 Mc
W8x10	30.8	7.8	15.5	4.2	10	26	17.6 LL	16.0 LL	14.9 LL	14.0 Mc	13.2 Mc	12.4 Mc	11.7 Mc	11.1 Mc	10.6 Mc	10.1 Mc
W6x16	32.1	10.2	20.2	4.3	22	45	17.9 LL	16.2 LL	15.1 LL	14.2 LL	13.5 LL	12.9 LL	12.4 LL	12.0 LL	11.6 LL	11.3 LL
W8x13	39.6	9.9	19.6	4.2	18	38	19.2 LL	17.4 LL	16.2 LL	15.2 LL	14.4 LL	13.8 LL	13.1 LL	12.4 Mc	11.9 Mc	11.4 Mc
W8x15	48.0	11.8	23.4	4.2	20	38	20.4 LL	18.6 LL	17.2 LL	16.2 LL	15.4 LL	14.7 LL	14.2 LL	13.6 Mc	12.9 Mc	12.4 Mc
W10x12	53.8	10.9	21.6	3.9	11	42	21.2 LL	19.3 LL	17.9 LL	16.8 LL	15.6 Mc	14.6 Mc	13.8 Mc	13.1 Mc	12.5 Mc	11.9 Mc
W8x18	61.9	15.2	30.1	5.5	18	29	22.2 LL	20.2 LL	18.7 LL	17.6 LL	16.8 LL	16.0 LL	15.4 LL	14.9 LL	14.4 LL	14.0 LL
W10x15	68.9	13.8	27.3	4.2	17	40	23.0 LL	20.9 LL	19.4 LL	18.3 LL	17.4 LL	16.4 Mc	15.5 Mc	14.7 Mc	14.0 Mc	13.4 Mc
W8x21	75.3	18.2	36.0	5.6	21	45	23.7 LL	21.6 LL	20.0 LL	18.8 LL	17.9 LL	17.1 LL	16.5 LL	15.9 LL	15.4 LL	15.0 LL
W10x17	81.9	16.2	32.1	4.2	18	45	24.4 LL	22.2 LL	20.6 LL	19.4 LL	18.4 LL	17.6 LL	17.0 Mc	15.9 Mc	15.2 Mc	14.5 Mc
W8x24	82.8	20.9	41.4	6.9	21	46	24.5 LL	22.2 LL	20.7 LL	19.4 LL	18.5 LL	17.7 LL	17.0 LL	16.4 LL	15.9 LL	15.4 LL
W12x14	88.6	14.9	29.5	3.5	12	30	25.0 LL	22.8 LL	21.1 LL	19.6 Mc	18.2 Mc	17.0 Mc	16.1 Mc	15.3 Mc	14.6 Mc	13.9 Mc
W10x19	96.3	18.8	37.2	4.2	20	45	25.7 LL	23.4 LL	21.7 LL	20.4 LL	19.4 LL	18.6 LL	17.9 LL	17.1 Mc	16.3 Mc	15.6 Mc
W8x28	98.0	24.3	48.1	6.9	20	55	25.9 LL	23.5 LL	21.8 LL	20.6 LL	19.5 LL	18.7 LL	18.0 LL	17.3 LL	16.8 LL	16.3 LL
W12x16	103.0	17.1	33.9	4.1	14	36	26.3 LL	23.9 LL	22.2 LL	20.9 LL	19.5 Mc	18.2 Mc	17.2 Mc	16.3 Mc	15.6 Mc	14.9 Mc
W10x22	118.0	23.2	45.9	6.1	18	41	27.6 LL	25.0 LL	23.2 LL	21.9 LL	20.8 LL	19.9 LL	19.1 LL	18.5 LL	17.9 LL	17.3 Mc
W12x19	130.0	21.3	42.2	4.2	17	40	28.5 LL	25.9 LL	24.0 LL	22.6 LL	21.5 LL	20.3 LL	19.2 Mc	18.2 Mc	17.4 Mc	16.6 Mc
W10x26	144.0	25.4	50.3	6.1	22	49	29.4 LL	26.8 LL	24.8 LL	23.4 LL	22.2 LL	21.2 LL	20.4 LL	19.8 LL	19.2 LL	18.6 LL
W12x22	156.0	27.9	55.2	4.3	21	49	30.2 LL	27.5 LL	25.5 LL	24.0 LL	22.8 LL	21.8 LL	20.9 LL	19.8 Mc	18.9 Mc	18.1 Mc
W10x30	170.0	32.4	64.2	6.1	29	58	31.1 LL	28.3 LL	26.2 LL	24.7 LL	23.5 LL	22.4 LL	21.6 LL	20.8 LL	20.2 LL	19.6 LL
W14x22	199.0	29.0	57.4	5.3	18	37	32.8 LL	29.8 LL	27.7 LL	26.0 LL	24.7 LL	23.6 LL	22.3 Mc	21.2 Mc	20.2 Mc	19.4 Mc
W14x26	245.0	35.3	69.9	5.3	20	46	35.1 LL	31.9 LL	29.6 LL	27.9 LL	26.5 LL	25.3 LL	24.4 LL	23.3 LL	22.3 Mc	21.4 Mc

L = Span from center to center of supports, ft. L must be > = dimension C, D, or E, whichever applies for single span beams. Base size of a continuous beam on the maximum of C, D, or E.
I = Moment of Inertia, in.⁴   S = Elastic Section Modulus, in.³
Mc = Allowable moment assuming Fb = 0.66Fy in accordance with the 1989 AISC ASD Specification, k-ft.
Lc = Maximum unbraced length of the beam in order to use this table and Mc.
RE = Max. beam end reaction for 1-3/4" bearing, kips.   RI = Continuous beam max. reaction at interior supports with 3-1/2" bearing, kips.
* No live load reductions have been included.   ♦ DL is in addition to beam weight.
Span dimensions are governed by either moment capacity (Mc), live load deflection of L/360 (LL), or interior (RI) or exterior (RE) bearing requirements.¹
Greater bearing dimensions are usually required for beams on non-steel supports.

A

FIGURE 5.30 (A) Partition of beam and (B) column tables.

Residential Steel Column Load/Spacing Tables - Pipe and Tube Columns

MAXIMUM COLUMN SPACING, when C = D or MAXIMUM TRIBUTARY LENGTH, (C+D)/2, when C ≠ D; ft.

TWO FLOORS (no roof or attic loads) - Unbraced Length of Column = 8 feet

		DL (psf) +	LL (psf)^
	1st flr	10	40
	2nd flr	10	30

COLUMN	Column Properties					TRIBUTARY WIDTH SUPPORTED BY THE CENTER BEAM - (A+B)/2								
SIZE	Weight/Ft	Fy	A	Pe	Pa	8'-0	10'-0	12'-0	14'-0	16'-0	18'-0	20'-0	22'-0	24'-0
3'dia. STD.	7.58	36	2.23	16	34	19.8	16.1	13.6	11.8	10.4	9.3	8.4	7.7	7.0
TS 3x3x0.1875	6.87	46	2.02	17	35	21.4	17.5	14.8	12.8	11.3	10.1	9.1	8.3	7.6
3.5'dia. STD.	9.11	36	2.68	22	44	27.2	22.2	18.7	16.2	14.3	12.8	11.5	10.5	9.7
TS 3x3x0.2500	8.81	46	2.59	21	44	26.5	21.6	18.2	15.8	13.9	12.4	11.2	10.3	9.4
3'dia. X-Strg.	10.25	36	3.02	20	45	25.7	20.9	17.7	15.3	13.5	12.0	10.9	9.9	9.1
TS 3x3x0.3125	10.58	46	3.11	24	51	30.3	24.7	20.8	18.0	15.9	14.2	12.8	11.7	10.8
4'dia. STD.	10.79	36	3.17	28	54	35.0	28.9	24.4	21.1	18.6	16.6	15.0	13.7	12.6
TS 4x4x0.1875	9.42	46	2.77	32	58	35.0	32.9	27.8	24.0	21.2	18.9	17.1	15.6	14.4
3.5'dia. X-Strg.	12.50	36	3.68	29	59	35.0	29.3	24.7	21.4	18.8	16.8	15.2	13.9	12.8
TS 4x4x0.2500	12.21	46	3.59	41	75	35.0	35.0	35.0	30.6	27.0	24.1	21.8	19.9	18.3
4'dia. X-Strg.	14.98	36	4.41	38	75	35.0	35.0	33.2	28.7	25.3	22.6	20.5	18.7	17.2
3'dia. XX-Strg.	18.58	36	5.47	33	77	35.0	33.6	28.4	24.5	21.6	19.3	17.5	15.9	14.7
5'dia. STD.	14.62	36	4.30	45	78	35.0	35.0	35.0	33.7	29.7	26.5	24.0	21.9	20.1
TS 4x4x0.3125	14.83	46	4.36	48	90	35.0	35.0	35.0	35.0	31.8	28.4	25.7	23.4	21.6

Column loads are based on a maximum eccentricity of 1" between the resultant (total) load and the centerline of the column.

Supported beams may be single span or continuous with a maximum eccentricity of 1" for the resultant load.   K = 1.0

F_y = Minimum design yield stress per the AISC Specification, ksi.

A = Gross cross-sectional area of column per the AISC Manual, in.

P_e = Maximum axial load with an eccentricity of 1", per the AISC Manual, kips.

P_a = Allowable axial load values

* No live load reductions have been included.   + DL is in addition to beam weight & 50psf for the interior walls.

(C + D)/2 has been limited to 35 feet to correspond with the beam tables.

Column bearing design must be per the AISC Specification.

B

FIGURE 5.30 (Continued)

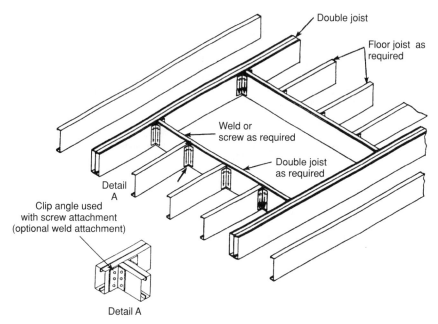

Double joist

Floor joist as required

Weld or screw as required

Double joist as required

Detail A

Clip angle used with screw attachment (optional weld attachment)

Detail A

FIGURE 5.31   Typical framing around floor opening.

FRAMING STAIRWELLS

First-floor-to-basement stairwells are below the main frames and can be installed parallel to the floor joists (Figure 5.33). As for most floor openings, install a second joist next to each side of the stairwell to act as a header. Assemble the entire stairwell with 3 × 3 angle that has been field cut into 6-inch-long pieces from 10-foot stock lengths. Install with #12, 14-×-¾-inch self-drilling screws. Cap the ends of the stairwell with field-cut pieces of rim track, installed with #8, 18-×-½-inch panhead self-drilling screws.

A typical assembly sequence for a first-floor-stairwell framing procedure follows.

- Field locate the leading tie beam, which is located at the head of the stairs, as required by the location of the stair landing. Field burn or drill ¾-inch-diameter holes in both the main-frame floor beams. Install the tie-beam with ⅝-inch-diameter bolts and the provided backing plates.

- Field locate the trailing tie-beam 15 feet away from the leading tie-beam. Install using the same procedure as for the leading tie-beam.

- Field locate both 10-inch-×-15-foot-long main floor beams as required by the stairwell location and rough opening size. Install following the procedure used for the tie-beams.

- The rough opening length of the stairwell is now 15 feet. The rough opening can be shortened to the desired length by spanning floor joists across the stairwell at the trailing end.

- Note that each time a floor joist bears across a floor beam, it must have a joist stiffener or lap another joist. This provides the additional required stiffness at bearing points and cannot be omitted. Install the stiffener with six #12, 14-×-¾-inch-long self-drilling screws.

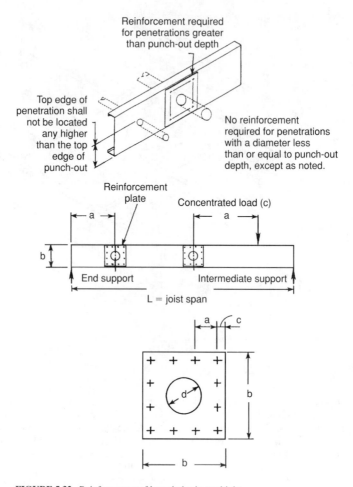

FIGURE 5.32 Reinforcement of large holes in steel joist.

- Install 8-inch rim track with #8, 18-×-½-inch panhead self-drilling screws to complete the rough stairwell opening.

BRIDGING THE FLOOR JOISTS

Floor joists with long, unsupported spans are likely to twist, sway from side to side, bounce up and down, or vibrate under live loads on the floor they support. To stiffen the floor system, reduce joist movement, and spread floor loads over more than one joist, install braces between the joists. This is called *bridging*. Three types of bridging commonly are used in lightweight steel-joist framing.

X Bridging. This bridging, also called *diagonal* or *herringbone* bridging, is set in pairs between the joists and then fastened with screws or welds (Figure 5.34).

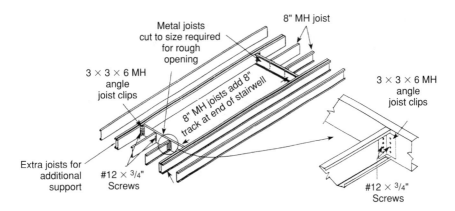

FIGURE 5.33 First-floor stairwell framing.

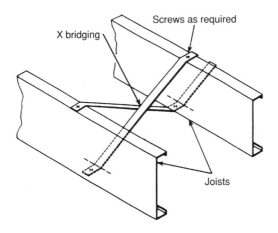

FIGURE 5.34 Typical X bridging installation.

Solid Blocking. Blocking might be a section of stud, joist, or track that is the full depth of the framing member being blocked. Tightly fit the section between the members. Anchor the blocking at each end (Figure 5.35). Solid blocking can be used in lieu of web stiffeners.

Flat-Strap and Block Bridging. Often called *lateral floor bracing*, this method uses a solid block with an 18-gauge steel strap (Figure 5.36). Place the bridging diagonally between opposite flanges of the framing members or run from flange to flange along the same face of the framing members. Anchor the bridging to each framing member. Space floor-joist bridging as noted in Table 5.2, except when the design requires or can accommodate an alternate spacing.

INSTALLING THE SUBFLOORING

Place the subflooring over the joists. The subflooring serves as a platform for the first floor (Figure 5.37) and supports the subsequent framing as well as the finished floor. Plywood

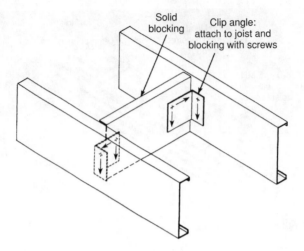

FIGURE 5.35 Typical solid blocking.

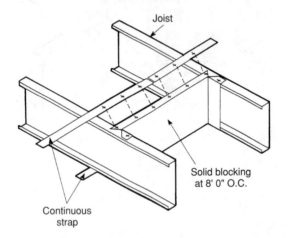

FIGURE 5.36 Typical flat-strap and block bridging.

TABLE 5.2 Recommended
Minimum Number of Rows of
Bridging for Floor Joists

Span, feet	Number of rows
Up to 14	1 row at midspan
14 to 20	2 rows at ⅓ points
20 to 26	3 rows at ¼ points

FIGURE 5.37   Subflooring installed over joists.

is the most commonly used subflooring material for residential building construction, although some builders and home manufacturers use particleboard, which is a thicker, but lighter sheet material made of wood fibers.

Lay the sheet with the long dimensions across the joists and stagger the ends of adjacent sheets. The partial sheets used to begin every other row can be ½, ⅔, or ⅓ of a sheet, depending on the overall dimension of the floor (Figure 5.38).

Plywood for subflooring comes in sheets that are nominally 48 × 48 inches, but often measure slightly less to allow for expansion. Edge treatment and thicknesses vary. The minimum thicknesses over typical steel joist spacings are:

- ½ inch with joists spaced on 16-inch centers
- ⅝ inch with joists spaced on 20-inch centers
- ¾ inch with joists spaced on 24-inch centers

Plywood is produced with its long edges squared, shiplapped, or finished with tongue-and-groove. Table 5.3 gives the recommended plywood grades for the underlayment sheets.

Install plywood underlayment only on a dry surface. Moisture, which can accumulate when the subflooring is exposed to weather during construction, might cause the dry underlayment panels to expand if the subflooring is not allowed to dry completely. A damp subfloor also can contribute to nail pops and squeaks. Normal scheduling, however, usually permits the subfloor to dry out and become conditioned in an enclosed, evenly heated environment before the plywood underlayment and floor covering are installed.

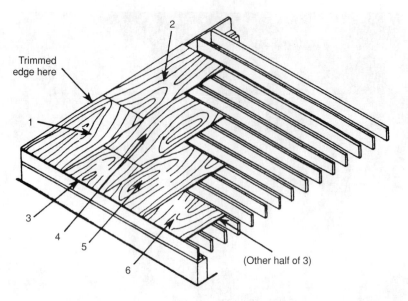

Trimmed
edge here

2

1

3

4

5

6

(Other half of 3)

FIGURE 5.38 Method of laying subflooring. Lay panels in order designated.

To avoid callbacks, inspect the subflooring surface for unevenness and flatness before installing the plywood underlayment. Uneven floor surfaces become obvious when smooth or shiny resilient floor covering is installed, especially in large areas that have strong sidelighting from windows, doors, or interior lighting. Observe the following precautions.

- When the subflooring panels are dry, and prior to installing the plywood underlayment, visually check the subflooring end and edge joints for evenness or variations in panel thickness that might telegraph through the underlayment. Use a short straight edge that is 12 to 24 inches long as a quick reference. If necessary, sand the subflooring joints with a commercial floor sander to smooth surfaces in the vicinity of the joints.

- Visually check the subflooring surface for flatness between the floor-framing members. Add blocking or plywood cleats under the floor, and fasten the subflooring to them with screws or nails to flatten the panels.

- Check the subflooring for squeaks and refasten the subflooring material as necessary before installing the underlayment. Abnormally low moisture content in the wooden floor also can contribute to nail pops and floor squeaks. When furnace or hot-air ducts are located in close proximity to the underside of the floor, the underfloor space should be well ventilated or insulated above the ducts in the joist cavity to keep the wooden floor from overdrying.

- Edge treatment and edge support depend on the type of finished flooring material that is to be applied over it. Wooden strip flooring is a structural material itself and requires no special support. Most other flooring materials, however, are laid on the underlayment sheets. Underlayment is not applied until the structure is completely enclosed. Subflooring with squared edges provides adequate support for both the structural flooring and underlayment.

TABLE 5.3. Recommended Plywood Grades for Underlayment

GRADE[1],[2]	EXPOSURE DURABILITY CLASSIFICATION	LOOK FOR THESE SPECIAL NOTATIONS IN PANEL TRADEMARK[3]	TYPICAL TRADEMARKS
APA Underlayment	Exposure 1	Sanded Face	
APA C-C Plugged APA Underlayment C-C Plugged	Exterior	Sanded Face	
APA A-C APA B-C APA A-D APA B-D	Exterior Exterior Exposure 1 Exposure 1	Plugged Crossbands Under Face[4]	
APA Underlayment A-C APA Underlayment B-C	Exterior	Sanded Face	

(1) *Veneer-faced, 19/32-inch or thicker panels of APA Rated Sturd-I-Floor, Exposure 1 or Exterior marked "Sanded Face," or APA Marine Exterior plywood also may be used for underlayment under vinyl or other resilient floor covering.*
(2) *Specific plywood grades and thicknesses may be in limited supply in some areas. Check with your supplier before specifying.*
(3) *Recommended for use under resilient floor covering.*
(4) *"Plugged crossbands (or core)," "plugged inner plies" or "meets underlayment requirements" may be indicated as alternate designation in or near trademarks.*

• If a nonstructural flooring material, such as carpeting, is laid directly on the subfloor, then the long edges of the plywood must be either shiplapped or tongue-and-groove. If square-edged plywood is employed, support the edges on blocking between joists.

It takes a specially designed screw to put down plywood on steel joists. The screw has threads, but they start about ¾ of an inch up from the drill point. The theory behind this special screw is that when the spinning screw goes through the plywood, it does not climb

FIGURE 5.39 An automatic screwgun with an extension handle makes installing the subflooring and underlayment easier.

the drive. Instead, it goes to the middle of the screw, penetrates the steel, and then the screw threads engage in the metal for a tight fit. As shown in Figure 5.39, there are automatic screwguns with an extension handle that makes driving the screws into the subflooring much easier.

Particleboard and OSB for the subflooring comes in 4-×-8-foot and 2-×-8-foot panels and is 1½ to 2 inches thick. Do not use OSB over joist spacings that are greater than 16 inches on center. Particleboard does not warp, but it does absorb moisture. Some codes do not permit its use as subflooring.

Available floor adhesives permit the use of single-layer floors instead of the conventional two-layer system. The advantages to such a system include reduced labor and material costs and the elimination of squeaking and popped nails. Floor stiffness also is increased by as much as 70 percent. When glued, the flooring and joists are fused into an integrated unit.

Use underlayment-grade tongue-and-groove plywood with a thickness of ½, ⅝, or ¾ inch. Run the outer plies, or *face grain*, perpendicular to the joists. Use ¹⁄₁₆-inch spacing at all joints. Apply a bead of glue with a caulking gun before covering each joist.

CHAPTER 6
LOAD- AND NONLOAD-BEARING WALLS

The walls of conventional steel-framed structures are either load- or nonload-bearing. Both interior and exterior walls can be load-bearing. Load-bearing walls support the weight of the building and are subject to horizontal wind or seismic forces as well as axial loads that are generated by roof and floor assemblies (Figure 6.1).

Special detailing must be given to the vertical alignment of studs and joists to allow for the proper transfer of loads throughout the structure. Openings in walls require headers designed to transfer joist and truss loads to jamb assemblies. Where bearing walls impose loads on floor members, attach stiffeners to the joist web. Check the web crippling at the ends of the headers and where any concentrated loads fall within the span of the headers.

Nonload-bearing walls do not support the structure above. Although interior walls can be load-bearing, most are nonload-bearing. Their main purpose is to support interior finished walls (Figure 6.2).

ORIENTING LOAD-BEARING STUDS

Before beginning to frame the load-bearing walls, make sure there is a print of the floor plan at the job site. To protect the print from weather and abuse, tape it to a piece of cardboard or thin plywood and cover it with plastic. Then clean the slab or floor area on which you are working. Remove all obstacles that might get in the way of snapped chalklines.

Begin the stud layout by snapping a chalkline on the inside of the main beam within the enclosed walls. Make sure the chalkline is in line with the inside edge of the beam. On the side or endwalls, check the plan to determine the location of these walls and then snap a chalkline at these locations. When you snap a chalkline that is more than 15 feet long, have a third worker snap from the center. The correct procedure is to have the third worker press his or her finger on the chalkline at the halfway point and then snap the chalkline on one side and then on the other.

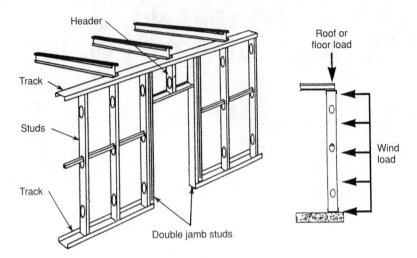

FIGURE 6.1 Exterior load-bearing walls showing axial and wind loads.

FIGURE 6.2 Interior walls can be load- or nonload-bearing depending on their purpose.

After there is a straight chalkline completely around the exterior of the home, start marking the floor between the columns (Figure 6.3). One worker can do this job by clamping the tape to the beam at one end. Leaving the tape measure in place, mark the floor with a carpenter's pencil using an arrowhead every 24 inches. The arrowheads mark the locations of the exterior studs. Leave the tape measure in place and you are less likely to make or compound an error. Do not worry about the doors and windows at this time.

On the endwalls, start from the outside of the beam and mark the layout 24 inches on center. Be sure the 24-inch-on-center marks are inside the chalkline so that the base tracks do not cover them. These 24-inch-on-center marks assure that the exterior sheathing can be installed with a minimum of cuts.

After you mark the exterior stud locations, check the plans for the location of all doors and windows. Identify the doors and mark their locations on the floor with a lumber crayon. Before you start any window or door layout, determine the correct rough opening required by the door and window manufacturers. If you frame out the doors and/or windows with the wrong rough-opening dimensions and then put on the exterior sheathing, it is time-consuming to make a change. Spend a little time obtaining the information before you frame the exterior and save a lot of time and effort later.

NOTE: The dimensions shown on Figures 6.4 and 6.5 are for illustration purposes only. Do not use them as standards. Generally, the steel runners, or track, should not be installed until the foundation has been allowed to cure for at least seven days.

Study the rough-opening dimensions for windows and doors and note the additional marks placed on the floor and slab. The crayon arrowheads are marked at the rough-opening location. The X next to the arrowhead indicates the side on which the stud is to be placed. Leave the 24-inches-on-center exterior stud marks in place. Then when the window and door framing is complete, you can continue to install short pieces of stud, or *crip-*

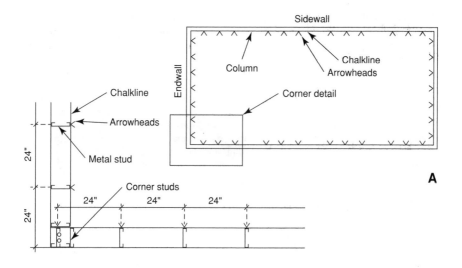

FIGURE 6.3 Layout of walls: (A) exterior stud locations and (B) corner details.

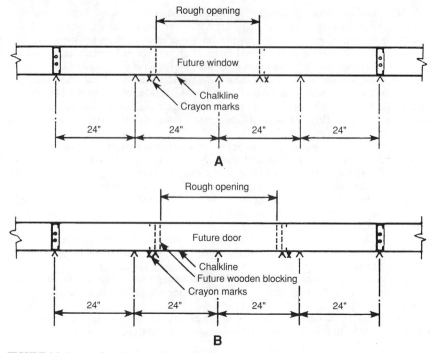

FIGURE 6.4 Layout of windows and doors: (A) mark window locations on the floor with a lumber crayon; (B) mark door locations the same way.

ples, following the 24-inches-on-center pattern for the sheathing and gypsum board. The door opening example shows the marked area where 2-× wooden material can be attached to the door-jamb steel studs. This makes it easier to square and plumb the door. Mark this on the floor along with the door size.

All studs normally are laid out with the open C turned in one direction. The only exception is at door and window openings, where a flat surface is needed on each side of the opening.

Cut the track for the bottom runner and top sill the same length, except at the doors. Do not install track across the bottom of the door openings. Save all scrap pieces of track for later use.

NOTE: The worker responsible for laying out the studs is one of the key workers on the job because the accuracy of the layout is important to the overall quality of the final job. This worker must be aware of the special framing requirements for all other trades such as plumbers, electricians, plasterers, drywallers, hangers, and millwork installers.

INSTALLING LOAD-BEARING STUDS

The steel framing can be assembled and then erected in place (Figure 6.6) or assembled stick-by-stick at the job site (Figure 6.7). In the latter procedure, install the track or runner for both the load-bearing and nonload-bearing walls in the same manner as the foundation perimeter track (see Chapter 5). Securely anchor the track or runner to the floor and overhead structure as indicated by the plans, specifications, and codes.

Squarely seat axial-loaded studs in the track (Figure 6.8). Abut the end of the stud to

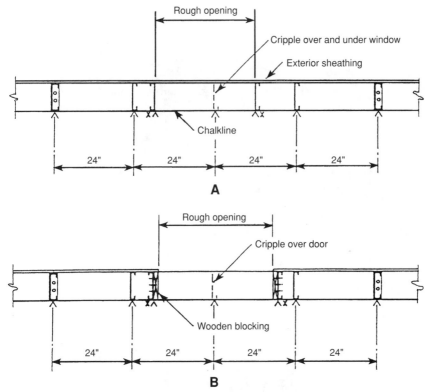

FIGURE 6.5 Stud orientation: (A) window opening and (B) door opening.

the track web. Plumb, align, and securely attach all studs to the flanges of both the upper and lower track by screws or welds. Position the studs in the track so that they are aligned directly below the floor, roof, or ceiling framing members above. If the studs are not aligned, provide headers.

Place ladders between the joists on 16-inch centers wherever the top runner cannot be attached directly to the joists (Figure 6.9). For first-floor systems, screw the track in place with one #6, 20-×-1¼-inch self-drilling screw at the floor-joist positions. Prior to installation, provide a bearing material (e.g., building paper, shims, caulk, or grout) between the underside of the bottom steel track and the top of the foundation to maintain a uniform bearing surface for the steel members. This also helps prevent moisture from seeping under the foundation and keeps out drafts, pests, etc.

On the sidewalls, attach pieces of scrap stud to each side of the columns before you install the top plate. Lean a stud against the column and make a mark ¼ inch above it. Do this on each side of all columns. This is the top mark for all of the scrap stud pieces. Attach the scrap pieces to the column with ¾-inch purlin-to-structure screws or with welds. Install the top track between the columns and attach with #8, ½-inch self-drilling panhead screws. After the top track is in place, the sidewalls are ready for stud installation.

For most models, the ceiling furring channels are installed on the endwalls first because the top track attaches to this furring. To align the top track with the bottom, use a stud slightly longer than the floor to ceiling height and a 4-foot level. Place one end of the

FIGURE 6.6 Steel framing can be assembled and then erected in place.

FIGURE 6.7 Steel framing assembled by the stick-by-stick method.

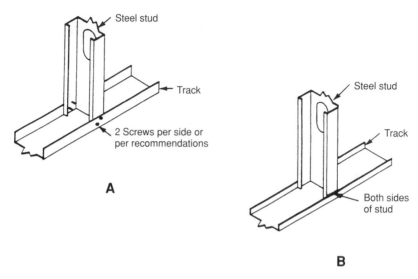

FIGURE 6.8 Axial-loaded stud must sit squarely in the runner.

FIGURE 6.9 Place blocks between joists on 16-inch centers wherever the top runner cannot be attached directly to the joists.

stud against the lower track and put the level on the stud. Move the stud in and/or out until it is plumb. Then put the top track against the stud and clamp it to a furring channel. Repeat the process at the other end. Once both ends are clamped, attach the track to the furring with ½-inch metal panhead screws.

When all the track is in place, install the studs. Start with the endwalls as they usually require cutting for sloping ceilings. Begin with the areas that require cut studs and you can use the short end pieces for cripples over and under the windows, over the doors, or for securing track.

Secure the bottom of the stud at the layout marks. Make sure that the stud is positioned squarely, or perpendicular, to the track. Use a square or laser. After the bottom is attached, put a 4-foot level on the stud, plumb it, and then attach it to the top track. At stud-wall corners, provide a minimum of three studs, located so as to provide surfaces on which all interior and exterior facings can be attached (Figure 6.10). Provide jackstuds between all the track and window sills, between window and door headers and the top track, at free-stand-

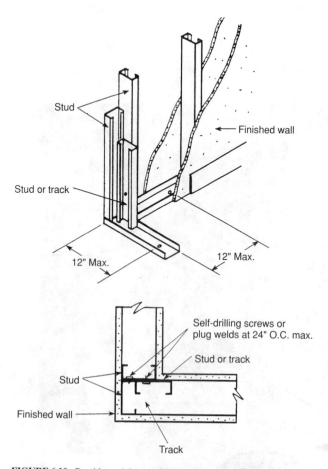

FIGURE 6.10 Provide a minimum of three studs at the corners of stud walls.

ing stair rails, and elsewhere to furnish structural support. Securely attach them to the supporting members.

Framing Intersections and Openings

At stud-wall intersections, you usually need a minimum of four steel studs (Figure 6.11). Remember that the load-bearing walls with steel studs and track must have a continuous, uniform base support. Less-than-full-height, or cripple, studs that are installed between an opening header and the bearing elevation of the members above should be designed to transfer all axial loads from the members above to the header. Squarely seat these cripple members against the webs of the track.

Bearing surfaces for joists, rafters, trusses, and the bottom track of axial-load-bearing walls should be uniform and level to assure full contact with the bearing flange or track web on the support over the required bearing and anchorage area. The maximum gap between the end of the stud and the web of the track should not exceed 0.063 inch. Vertically align all axial-load-bearing studs and transfer all loads to structural supports or foundations. Maintain this vertical alignment at floor-wall intersections.

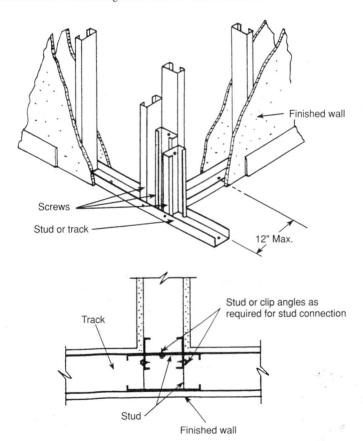

FIGURE 6.11 Provide a minimum of four studs at the intersection of stud walls.

TABLE 6.1 Maximum Allowable Axial Loads for 3⅝-Inch Studs, KIPS (1000 Pounds) Per Stud

			3⅝" stud designation			
Lateral pressure	Wall height	Stud spacing	20 ga.	18 ga.	16 ga.	14 ga.
0 psf	8'	NA	2.32	3.14	4.78	6.19
(No wind)	9'	NA	2.24	3.04	4.48	5.77
	10'	NA	2.15	2.92	4.14	5.28
20 psf	8'	12" O.C.	2.05	3.07	5.31	7.11
.		16"	1.75	2.74	4.99	6.76
		24"	1.18+	2.12+	4.38	6.10
	9'	12" O.C.	1.71+	2.66	4.66	6.28
		16"	1.35+	2.26+	4.28	5.87
		24"	0.71*	1.55+	3.58+	5.10+
	10'	12" O.C.	1.36+	2.22+	3.98	5.39
		16"	0.97*	1.78+	3.56+	4.94
		24"	—	1.01*	2.82*	4.12+
25 psf	8'	12" O.C.	1.82	2.82	5.07	6.95
		16"	1.46+	2.42	4.68	6.42
		24"	0.79*	1.68+	3.95+	5.63
	9'	12" O.C.	1.44+	2.36	4.37	5.97
		16"	1.02*	1.90+	3.92+	5.47
		24"	—	1.06*	3.10+	4.57+
	10'	12" O.C.	1.06*	1.89+	3.66+	5.05+
		16"	0.61*	1.38*	3.18+	4.52+
		24"	—	—	2.31*	3.56

NOTES:
  Studs shall be braced against rotation. Install bridging spaced at intervals not to exceed 48" on center. Stud shall be attached to continuous track.
  Where values are omitted, stud deflection exceeds L/240 at specified lateral load. Values followed by (*) do not exceed L/240; by (+) exceed L/360.

The exception to this is where load-carrying members do not align with a load-distributing member. Provide for the transfer of loads from joist-bearing to axial-load-bearing studs. Members can be added between the members, which have been placed at the specified spacing, to support members that are not in alignment. These added members should transfer loads on a continuous load path to a foundation or structural support. Table 6.1 gives the maximum axial loads for typical 3⅝-inch steel studs.

When the continuity of axial-load-bearing walls cannot be maintained because of an interruption by nonaligned floor joists, place short sections of joists, studs, track, or additional joist members that are capable of transferring the loads in alignment with the stud above. As an alternative, a small section of stud, called a *filler*, with an axial capacity at least equal to the capacity of the stud above can be used to transfer the loads.

Splicing and Coping Studs

Splicing together members is not recommended because the splice can alter the integrity of the material. Using shears to remove portions of the flanges, or *coping*, also is not rec-

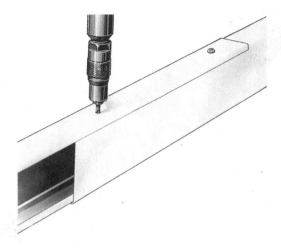

FIGURE 6.12 When splicing two studs, nest one into the other to form a box section.

ommended because it can lessen the bearing strength of the stud. Reinforce areas where splicing or coping is required. Figure 6.12 shows how a stud can be spliced together. To splice two studs, nest one into the other and form a box section that is at least 8 inches deep. Fasten the studs together with two ⅜-inch type-S panhead screws in each flange. Locate each screw no more than 1 inch from the ends of the splice. Figure 6.13 shows how to splice together a track.

LAYING OUT EXTERIOR WALL-FLOOR JUNCTIONS

Several perimeter foundation arrangements were given in Chapter 5. These plans all had the track laid on a sheathing deck, but did not indicate the location of the studs. In Figure

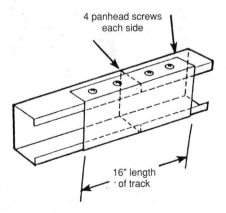

4 panhead screws each side

16" length of track

FIGURE 6.13 Splicing a track.

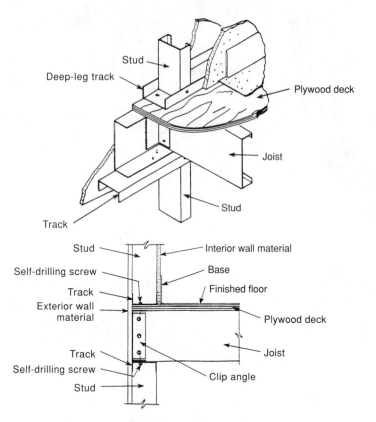

FIGURE 6.14 Exterior wall/floor junction details.

6.14, the stud is shown in the wall-floor junction detail. The methods for locating and fastening the steel in this figure are the same as those described in Chapter 5.

Figure 6.15 shows a version of a load-bearing exterior wall, a nonload-bearing exterior wall, and a load-bearing interior wall. While these illustrations show steel and concrete decking, plywood and any finished flooring material, including wood, tile, vinyl, and carpet, can be substituted for these materials. Load-bearing interior steel-framed walls can be anchored as detailed in Figure 6.16 in reinforced concrete or concrete-block foundations.

FRAMING DOORS AND WINDOWS

Steel accepts any kind of door (flush, panel wood, metal skin) or window (double-hung, casement, awning) on the market. The height of each opening usually is determined by the manufacturer's dimensions. The top rough openings of doors and windows generally are located at the same height, which is usually 83 inches above the subflooring.

For windows, check the rough-opening dimensions provided by the manufacturer. Measure down from the top to find the proper location for the bottom sill. Unlike con-

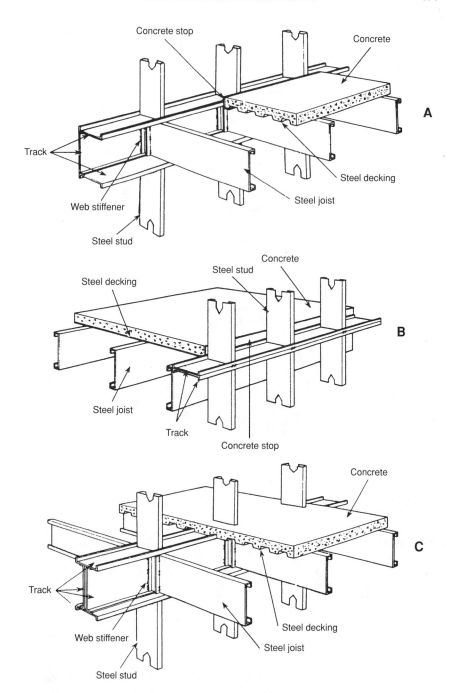

FIGURE 6.15 (A) Load-bearing exterior wall; (B) nonload-bearing exterior wall; (C) load-bearing interior walls.

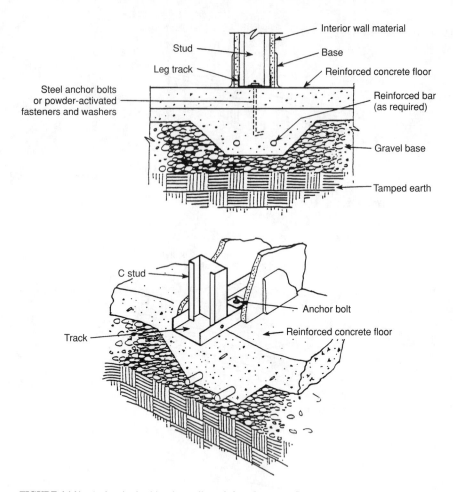

FIGURE 6.16A   An interior load-bearing wall to reinforced concrete floor.

structing with wood, extra headers are not needed over windows and doors. After the top plates and bottom sills are in place, measure and install the cripples as required to keep the 24-inch-on-center stud spacing for sheathing and drywall materials.

In the wall elevation view for both one- and two-story residential structures, the stud surfaces are broken for doors and windows (Figure 6.17). Door and window openings, however, require special construction for heads, sills, supports, jamb mountings, lintels, and cripple studs (Figure 6.18). Rough-in frame doors and borrowed-light openings with steel studs and runners. Cut the members for the headers and sills at least 8 inches longer than the width of the opening. This provides extra metal at each stud that can be used as tabs to connect the cross members to the vertical studs. Use snips to cut the legs free from the web at the ends and then fold to form tabs.

When assembling the header, work from the outside of the wall because the walls are not the same thicknesses. Get in the habit of measuring at the same place each time. Vertically position floor-to-ceiling structural studs adjacent to the frames. Then securely

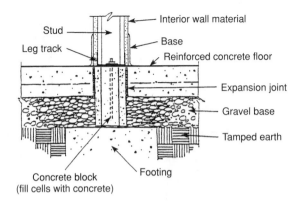

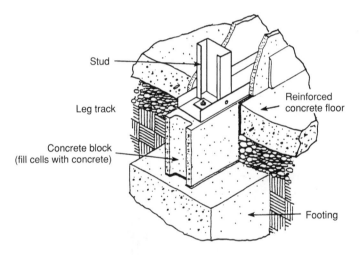

FIGURE 6.16B Interior load-bearing wall attachment to concrete block foundation.

anchor them to the top and bottom runners with metal lock fasteners or screws. Where heavy or oversized doors are used, install additional strut-studs at the jambs. Fabricate the sill and header sections from steel runners and install them over less-than-ceiling-height door frames and above and below the borrowed-light frames. From a section of runner, cut a piece approximately 8 inches longer than the rough opening, slit the flanges, and bend the web to allow the flanges to overlap the adjacent vertical strut-studs. Securely attach the studs with metal lock fasteners or screws.

For frames with jamb anchors (Figure 6.19), fasten the anchors to the strut-studs with two ⅜-inch, type-S-12 panhead screws. Install the cripple studs in the center above the door opening and above and below the borrowed-light openings spaced on maximum 24-inch centers. Attach wooden nailers directly to the studs if desired. Where control joints in the header boards are required, install cripple studs ½ inch away from the studs, but do not attach the cripple to the runners or studs.

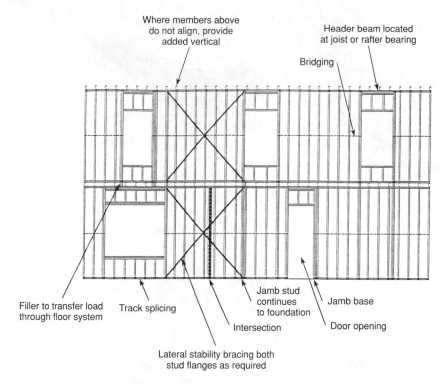

Where members above
do not align, provide
added vertical

Header beam located
at joist or rafter bearing

Bridging

Filler to transfer load
through floor system

Track splicing

Jamb stud
continues
to foundation

Jamb base

Intersection

Door opening

Lateral stability bracing both
stud flanges as required

FIGURE 6.17 One- and two-story elevation view of residential structure.

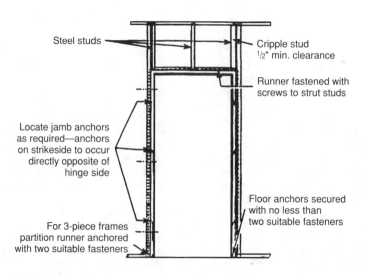

Steel studs

Cripple stud
$1/2$" min. clearance

Runner fastened with
screws to strut studs

Locate jamb anchors
as required—anchors
on strikeside to occur
directly opposite of
hinge side

Floor anchors secured
with no less than
two suitable fasteners

For 3-piece frames
partition runner anchored
with two suitable fasteners

FIGURE 6.18 Frames for standard doors.

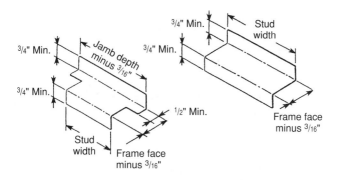

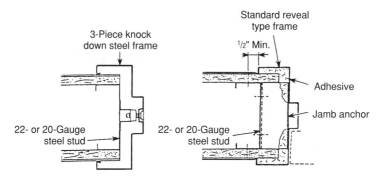

FIGURE 6.19 Typical jamb anchors.

Because door openings vary in size, the load must be considered. Figure 6.20 indicates the framing necessary for load-bearing and nonload-bearing walls with door openings less than 4 feet. Figure 6.21 details load-bearing and nonload-bearing walls with door openings greater than 4 feet. As shown in Figure 6.22, there are several types of header designs from which the framer can choose, including the structural angle header shown in Figure 6.23.

Jamb and header steel-framing installations (Figures 6.24 and 6.25) also depend on whether or not there is an axial load. The detailed drawings provide a full understanding of the jamb in steel-framed construction by taking the framer from the base of the framing (Figure 6.26), to the jamb at the bottom of the wall (Figure 6.27), to the jamb-stud header possibilities (Figure 6.28), to the jamb at the top of the wall (Figure 6.29). The framing details for the jamb and sill in either a door or window can be seen in Figure 6.30.

Installing Doors

While the framer usually does not install the door itself, he or she must have a knowledge of the many styles available:

- Single-unit steel door frames for single-swing doors not exceeding 4 feet in width
- Single-unit steel door frames for pairs of doors not exceeding 6 feet in width
- Three-unit slip-on door frames for single-swing doors not exceeding 3 feet in width

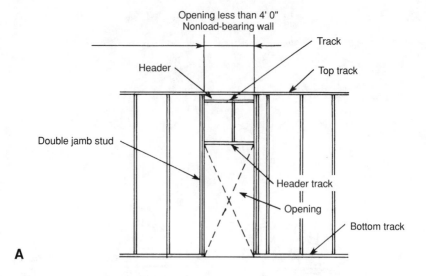

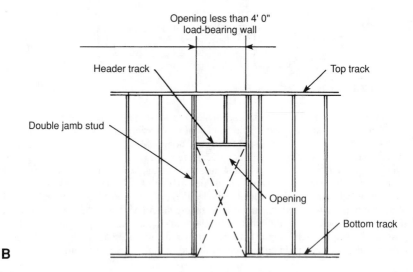

FIGURE 6.20 (A) Door openings less than 4 feet in load-bearing walls; (B) door openings less than 4 feet in nonload-bearing walls.

Framing Single-Unit Doors. For single-unit frames and standard-weight doors, the minimum recommended steel stud width is 2½ inches. For heavyweight doors and wide pairs of doors, the minimum recommended stud width is 3½ inches. The steel thickness of the studs that support the door frame should not be less than 20 gauge.

Install frames with steel anchors. Install anchors in the door header, similar to those in the side jambs, in frames for pairs of doors that are 5 feet or more in width. Locate the anchors 6 to 8 inches from the jambs and do not exceed 24-inch centers. Use the 8-inch frame

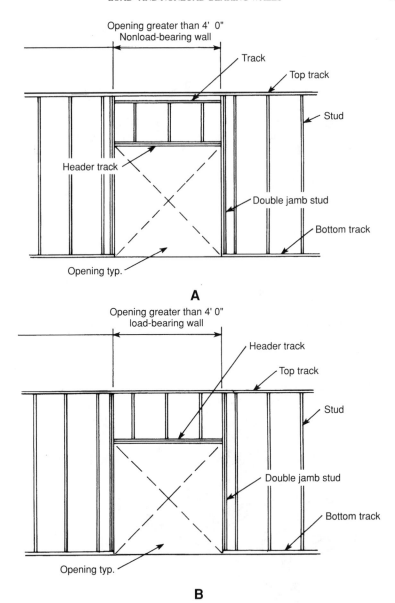

FIGURE 6.21   (A) Door openings greater than 4 feet in load-bearing walls; (B) door openings greater than 4 feet in nonload-bearing walls.

header to receive a door closer. These details are applicable for pairs of doors 5 feet or more in width and for all heavy doors.

Framing Three-Unit Slip-On Doors.   Three-unit slip-on door frames are designed for installation in a door opening after the steel-stud gypsum-board partition installation is complete.

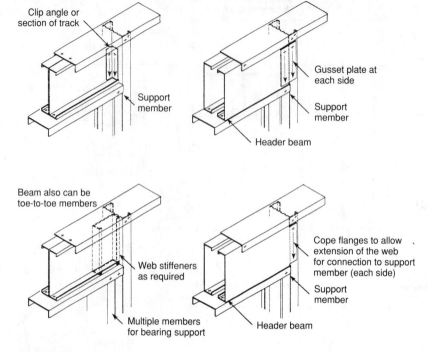

FIGURE 6.22 Various header designs.

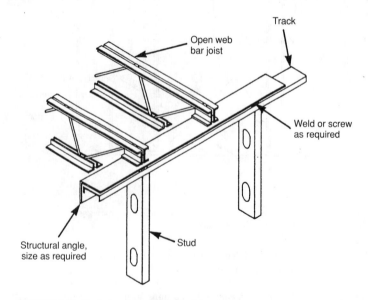

FIGURE 6.23 Typical structural angle header.

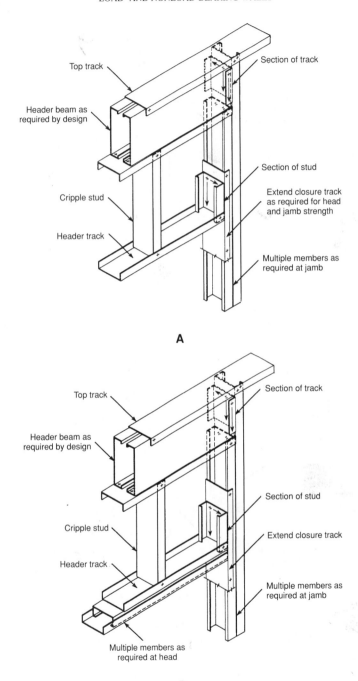

A

B

FIGURE 6.24 Axial load-bearing jamb and header.

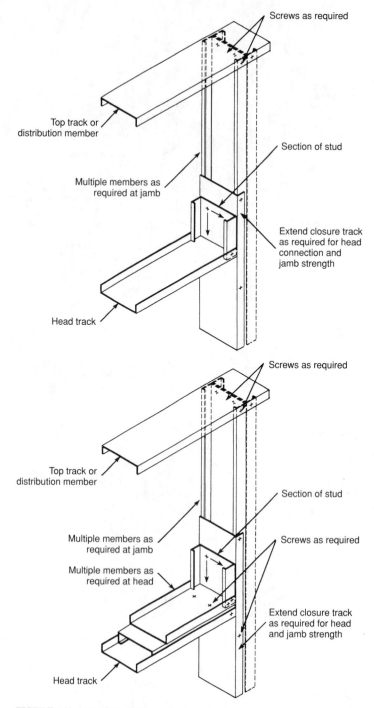

FIGURE 6.25 Nonaxial load-bearing jamb and head.

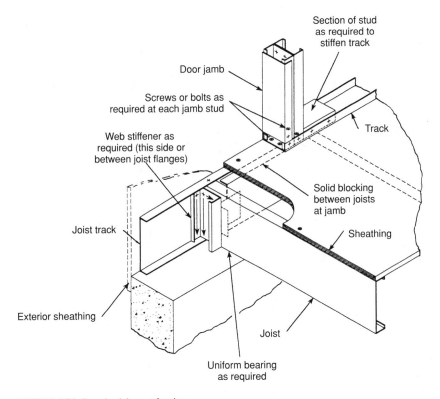

FIGURE 6.26 Door jamb base at framing.

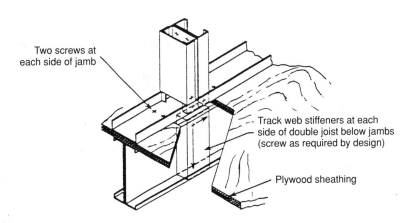

FIGURE 6.27 Jamb at bottom of the wall.

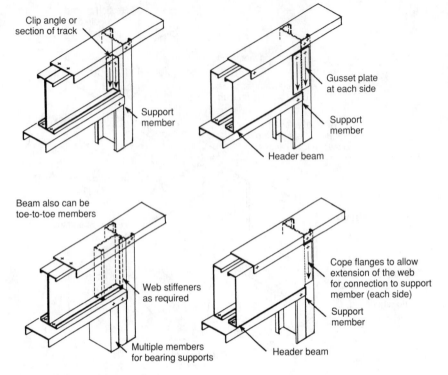

FIGURE 6.28 Header-to-jamb stud details.

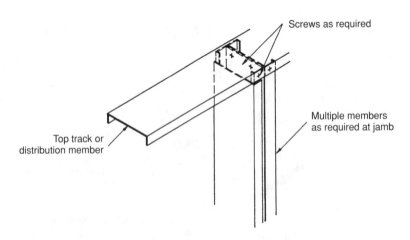

FIGURE 6.29 Jambs at top of wall.

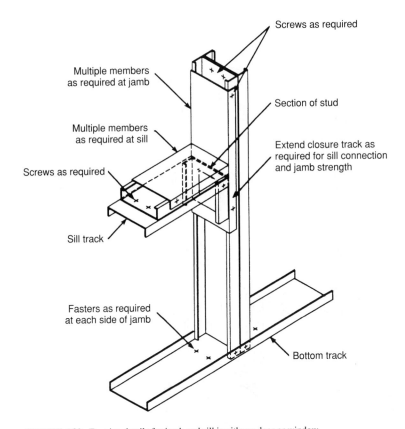

FIGURE 6.30 Framing details for jamb and sill in either a door or window.

Three-unit frames in fire-rated partitions should have a minimum thickness of 16 gauge. The frames should have double back-bend frame-face returns. The frame-jamb pieces should be designed for screw anchorage of the jambs to the partition frame (not shown), and the jamb and head pieces should be designed to provide a positive means of aligning and locking the jamb-head intersections in place.

Framing Windows

The nonaxial-load-bearing wall window opening is shown in Figure 6.31, along with its necessary framing. Like doors, window openings vary in size and have different framing requirements. Details for load- and nonload-bearing window openings less than 4 feet wide are given in Figure 6.32. Details for load- and nonload-bearing window openings greater than 4 feet wide are given in Figure 6.33. Figure 6.34 shows how to install a window sill in a panelized wall.

When installing metal window framing in climates that experience summer or winter temperature extremes that might result in condensation on the metal frames, isolate the gypsum base so that it does not come in direct contact with the frame. Provide protection

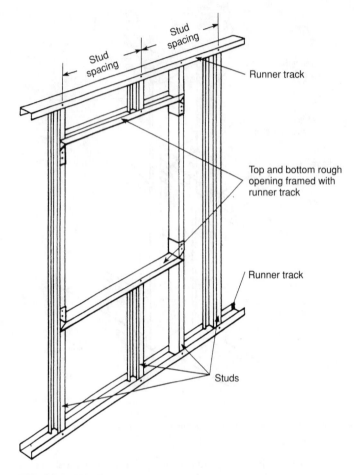

FIGURE 6.31 Wall window opening in a nonaxial load-bearing wall.

against moisture damage by placing metal trim between the gypsum-board base and the window frame. Separate the metal sash and metal trim and provide some measure of insulation between the two different metals by using a waterproof insulating tape that is ¼ inch thick and ½ inch wide or with a waterproof acrylic caulk. If an aluminum frame and steel trim come in direct contact in the presence of condensation, the moisture might cause electrolytic deterioration of the aluminum frame. A large wall opening, such as the one shown in Figure 6.35, basically is framed in the same manner as a smaller opening.

ATTACHING BRACING AND WALL BRIDGING

Once the assembled wall is erected, leveled, and squared, screw- or weld-attach a stud or runner diagonally across the studs until the sheathing at the corners or more permanent brac-

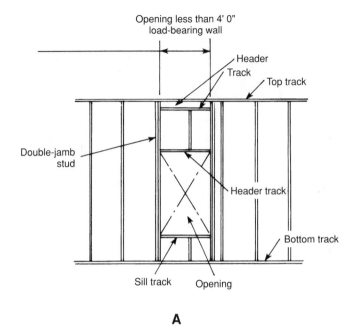

FIGURE 6.32 Window opening less than 4 feet wide in (A) load-bearing, and (B) nonload-bearing walls.

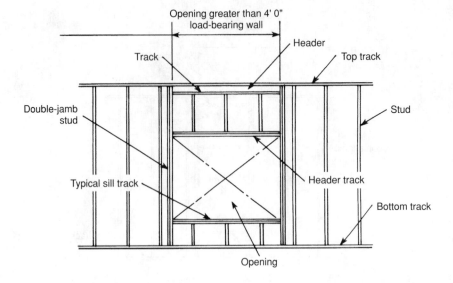

A

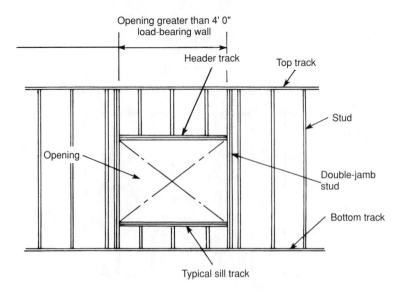

B

FIGURE 6.33 Window opening greater than 4 feet wide in (A) load-bearing, and (B) nonload-bearing walls.

FIGURE 6.34 Installation of a window sill in a panelized wall.

FIGURE 6.35 Frame large wall openings the same as small openings.

ing can be attached. Provide adequate bracing for all building systems until lateral stability systems, such as shear walls or braced frames, are installed and anchorage is completed.

Provide both lateral and diagonal bracing in accordance with the manufacturer's specifications or recommendations. Lateral bracing prevents the stud from bending about the minor axis due to axial loads and also stops flange rotation due to uniform wind loading (Figure 6.36). Sheathing and wallboard attached to both sides of the studs, if properly designed, can provide horizontal bracing and diaphragm strength. Solid blocking also can be used to provide lateral bracing (Figure 6.37).

During construction and before the wall sheathing is applied, floors and walls are loaded with construction loads. This load requires lateral bridging. Install horizontal bridging, when required, before overhead floor, roof, or ceiling structures are installed. Details of the possible bridging connections can be found in Figure 6.38. In this figure, the type-1 connection can be used with studs up to 6 inches, type-2 and type-3 connections can be used for any stud widths. The type-4 connection is screw-applied. As required, the joists must be bridged with horizontal steel strapping fastened to the bottom flange and secured to a positive restraint at the ends.

Use X bridging, which is similar to bridging used for joists or rafters, to reinforce the studs. This bridging is either a flat-strap, notched-channel, or proprietary system. Solid blocking also can be used to reinforce studs. When the thickness of the blocking material

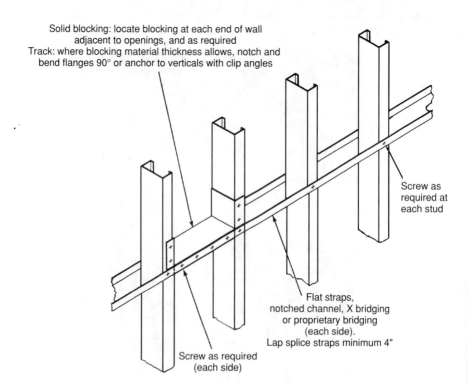

Solid blocking: locate blocking at each end of wall adjacent to openings, and as required
Track: where blocking material thickness allows, notch and bend flanges 90° or anchor to verticals with clip angles

Screw as required at each stud

Flat straps, notched channel, X bridging or proprietary bridging (each side). Lap splice straps minimum 4"

Screw as required (each side)

FIGURE 6.36 Lateral bracing prevents stud from bending and prevents flange rotation.

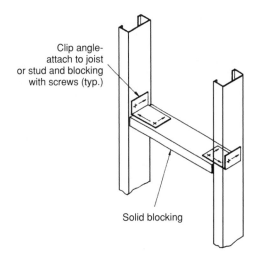

FIGURE 6.37 Using solid blocking to provide lateral bracing.

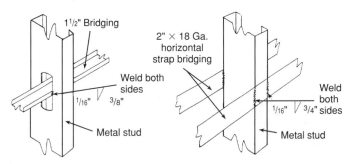

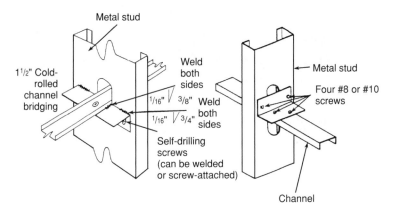

FIGURE 6.38 Various possible bridging connections.

allows, notch and bend the track 90 degrees for connection. When provisions are provided for the transfer of the flange forces to the solid blocking, the blocking need not be the full depth of the member.

Adequate bracing is recommended for both chords or member flanges until the sheathing is installed. When sheathing is applied to one side only, permanently install the required bracing on the unsheathed faces. If the sheathing is not designed to provide diaphragm strength, use diagonal bracing to resist racking loads imposed by winds or earthquakes. Diagonal bracing can be steel straps that are stressed in tension when resisting racking (Figure 6.39). Extend the straps from the bottom of one stud to the top of another, preferably at about a 45-degree angle from the horizontal. The angle at which the strap is installed significantly affects the strap's ability to resist racking loads.

As illustrated in Figure 6.40, securely fasten the ends of the horizontal strap bridging to the double-stud jambs at the openings and to the double-stud corner columns or other suitable rigid restraints. Screw or weld continuous steel straps horizontally to each flange of all the studs on the uncovered side(s) of the wall at midheight or use two continuous straps at third points. The spacing of and the number of straps required depends on wall height.

Figure 6.41 diagrams a method for obtaining diagonal stability by bracing at the floor transition. Figure 6.42 shows how bracing can be used to reinforce a two-story dwelling. Strap forces might require additional stiffening of the top and bottom track or structural angle. The design often requires a track distribution member for situations where joists and/or studs do not align with the studs below.

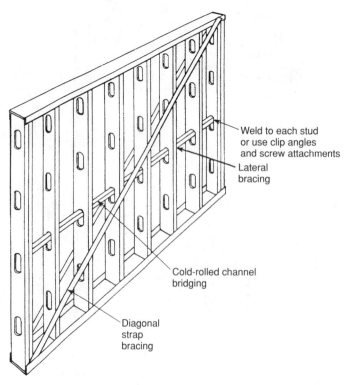

Weld to each stud
or use clip angles
and screw attachments

Lateral
bracing

Cold-rolled channel
bridging

Diagonal
strap
bracing

FIGURE 6.39 Typical diagonal-strap bracing.

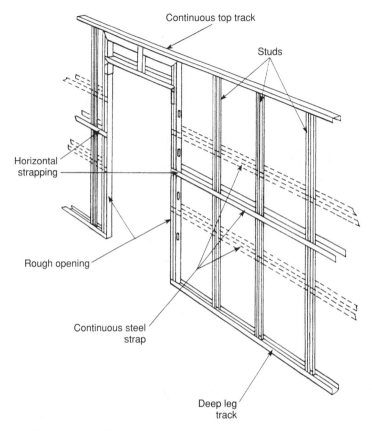

FIGURE 6.40 Typical horizontal-strap bridging.

During erection, use steel cables with turn buckles in sufficient number to prevent distortion and damage to the framework due to wind or erection forces. These cables also can be used to plumb and align the framework. Leave the cables in place to secure the lateral stability of the structure.

Framing the First-Floor System

The first-floor system generally includes the main support girder, floor joists, rim track, 2-inch-wide strap bracing, and required clips and fasteners. The first-floor system is different for straightwall- and slantwall-designed structures. For straightwall dwellings, the floor joists sit on top of the foundation wall. For slantwall buildings, the floor joists sit down between the foundation walls.

A typical assembly sequence for the first floor follows.

- Install support columns and girders, including temporary bracing.
- Erect the main frames and at least enough of the purlins (a minimum of four per bay) or supports to stabilize, plumb, and square the house (see Chapter 7).

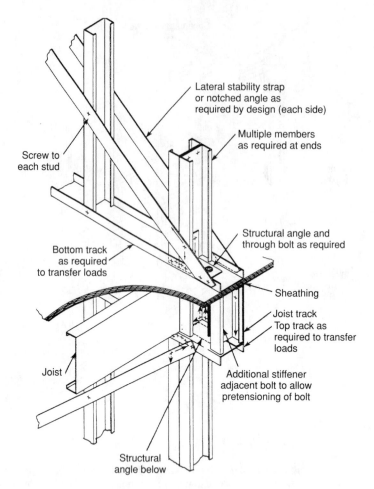

FIGURE 6.41 Diagonal stability bracing anchorage at floor transition.

- Install the inverted 8-inch track over the girder with #12, 24-×-1¼-inch-long self-drilling screws, staggered on 16-inch centers, but not aligned with the floor joists.
- Cut the rim track to include a tab for connection to each column.
- Lay out the floor joists on 16-inch centers per the floor specifications.
- Install the rim track on the outside of each bay section. This can be done easily by sliding the joists so that they overhang the wall and then using #8, 18-×-½-inch self-drilling screws at the top and bottom of each joist. Slide the joists back to their final positions and fasten the tabs to the columns.

   If the architectural design has bays with floor joists extending from outside wall to outside wall, temporarily bend the tabs out of the way so that you can slide the joist over the concrete wall on each side in order to install the bottom screw. Fasten the floor joists to the inverted track with #12, 14-×-¾-inch self-drilling screws. Lap floor joists approximately

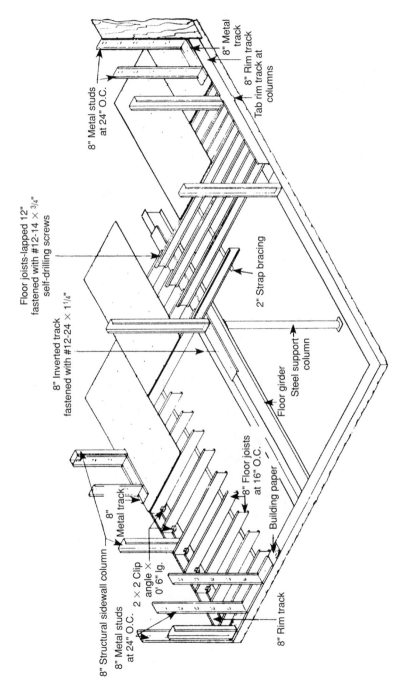

8" Metal track

8" Rim track
Tab rim track at columns

8" Metal studs at 24" O.C.

Floor joists-lapped 12" fastened with #12-14 × ³/₄" self-drilling screws

8" Inverted track fastened with #12-24 × 1¹/₄"

2" Strap bracing

Floor girder

Steel support column

8" Floor joists at 16" O.C.

Building paper

8" Rim track

8" Metal track

8" Structural sidewall column

8" Metal studs at 24" O.C.

2 × 2 Clip angle × 0' 6" lg.

FIGURE 6.42 Diagonal stability bracing on a two-story dwelling.

12 inches. Install six additional #12 screws to create a joist splice. This is required to resist the bearing forces at the girder. Field cut from the 10-foot stock lengths, 4- to 6-inch-long pieces of 2 × 2 angle. Fasten these clip angles to the floor joists with #12, 14-×-¾-inch self-drilling screws and then clip it to the concrete with powder-actuated nails.

Install 2-inch-wide, light-gauge strap bracing continuously across the bottom of the joists. The strap replaces the more common bridging used with wooden joists. Install the strap bracing perpendicular to the joists on 8-foot centers and fasten to each joist with one #12, 14-×-¾-inch-long self-drilling screw. Use two screws at the lap connections. Install the decking and notch around the frame columns.

Framing the Second-Floor Assembly

The second-floor framing system is the same geometric system used for the first floor (Figure 6.43). The floor joists are 8-inch C joists placed perpendicular to, and on top of, the floor beams. All floor joists are located on 16-inch centers. Locate rim tracks at each endwall to secure the floor joists and function in the floor system as a rim joist. Place the outside face of the rim joist flush or in line with the exterior wall below.

Lap the floor joists 12 inches and fasten at the lap with six #12, ¾-inch self-drilling, self-tapping screws. If joists are not lapped at one or more floor support beams, add a 12-inch piece of floor joist at each support beam to act as a joist stiffener.

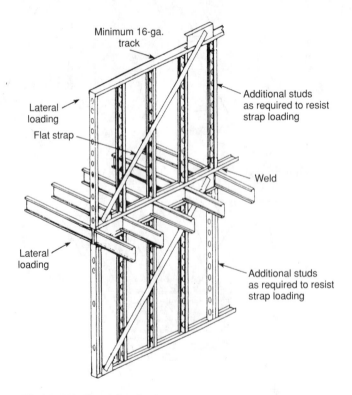

Minimum 16-ga. track

Lateral loading

Flat strap

Lateral loading

Additional studs as required to resist strap loading

Weld

Additional studs as required to resist strap loading

FIGURE 6.43 Second-floor framing system.

Secure the floor joists to the C-channel beams with #12, ¾-inch, self-drilling, self-tapping screws. Floor joists that are secured to steel wide-flange girders require an 8-inch inverted track that is fastened with #12, 24-×-1¼-inch fasteners staggered on 16-inch centers. Fasten the floor joists to the inverted track using #12, ¾-inch, self-tapping screws.

Fasten ¾-inch tongue-and-groove floor decking to the floor joists using #8, 18-×-1¹⁵⁄₁₆-inch flathead self-tapping screws on 12-inch centers. It is recommended that the decking be glued to the floor joists with a good adhesive. Also glue the tongue-and-groove joints. Leave approximately ⅛ inch between the endcuts.

Installing Shear-Wall Bracing

Wind and seismic forces cause a cyclical shearing action on a building that is known as *shear force* (see Chapter 2). To prevent damage from this force, stiffen the building frame with shear walls and/or crossbracing (Figure 6.44).

Provide stud walls at the locations indicated on the plans as *shear walls* for frame stability and lateral-load resistance. Then brace them as indicated on the plans and per the specifications. Provide additional studs per the plans and specifications to resist the vertical components. Prefabricated panels and subassemblies must be square and might require bracing to resist racking. Exercise care when lifting prefabricated panels and subassemblies to prevent local distortion.

Consult Table 6.2 for shear-bearing capacity considerations of the perimeter wall track. The table covers 16- to 25-gauge-thick steel with ⅛- and ³⁄₆₄-inch-diameter fasteners. For example, an 18-foot-high wall under a designed pressure requires ⅛-inch fasteners to be spaced on maximum 16-inch centers to avoid exceeding the designed shear stress of 25-gauge steel track. In addition, those same ⅛-inch-diameter fasteners, driven ¾-inch deep in 3000 pounds per square inch (psi) concrete, must be spaced on maximum 14-inch centers to avoid exceeding the designed shear stress of the concrete.

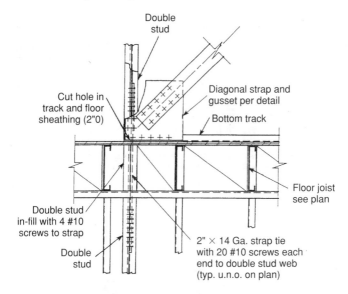

FIGURE 6.44 Shear walls and/or crossbracing.

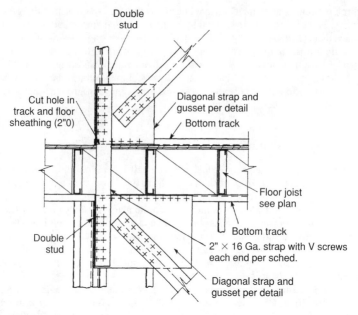

FIGURE 6.44 *(Continued)*

TABLE 6.2 Allowable Shear Load (in Pounds)

Shank diameter	Steel thickness				
	16 ga.	18 ga.	20 ga.	22 ga.	25 ga.
⅛"	—	—	310	259	181
%₄"	—	—	349	291	203
%₂"	—	—	388	323	226
¹¹⁄₆₄"	—	517	427	356	248
³⁄₁₆"	777	621	460	388	271

NOTE: It is important that the job engineer approve the type, size, and maximum spacing of perimeter fasteners to meet the design-load requirements.

Various shear-wall bracing designs are detailed in Figure 6.45. With any shear bracing, strap forces might be required for additional stiffening of the top and bottom track or structural angle. The stacked second-floor shear wall featured in Figure 6.46 uses one common gusset plate. The allowable shear for wind or seismic forces in psf for vertical diaphragms of steel stud-wall assemblies is given in Table 6.3.

TABLE 6.3 Allowable Shear for Wind or Seismic Forces for Vertical Diaphragms of Steel Stud Wall (in Pounds Per Foot) Assemblies

Max. stud spacing[1] (in.)	Material[3]			Screws		Nominal shear (plf)	Shear value (plf)[4]
	Type	Thickness (in.)	Minimum size	Maximum spacing[2] (in.)			
24	Gypsum	½	#6 × 1-inch	12		333	133[5]
16	Wallboard	½	#6 × 1-inch	12		425	170[5]
24	(both sides)	½	#6 × 1-inch	6-perimeter 12-interior		563	225[5]
		⅝ two ply	#6 × 1 (base) #6 × 1⅝ (face)	24-base ply 12-face ply		438	175[5]
24	Gypsum sheathing board (one side) gypsum wallboard (opposite side)	½	#6 × 1-inch	12		263	105[5]
24	CDX plywood (one side) gypsum wallboard (opposite side)	½	#6 × 1-inch	12		375	150
24	Expanded metal lath and portland cement plaster (3-coat) both sides	⅞	#8 × ½-inch (lath)	9-(lath)		450	180

[1] Walls shall be framed with a minimum 0.035-inch (20-gauge) structural C studs attached to a minimum 0.035-inch runner track by either #10 × ½-inch screws or by ⅛-inch-long fillet welds each side, top and bottom. Overall stud dimensions shall be a minimum 3½ × 1½ inches with a minimum yield strength of 33 ksi. Uplift of corners (end studs) shall be prevented by installation of clip angles and/or other connectors designed to attach wall to structural elements below.

[2] Applies to attachment at all studs and runner track. (Note: Shear values can be considered conservative for screw spacings less than those specified in this table. Building codes may dictate screw spacings less than the tested spacings).

[3] In addition to the requirements herein, comply with manufacturers requirements for installation.

[4] Shear values are the recommended design values based on the nominal shear capacity with a safety factor of 2.5.

[5] These values shall be reduced 50 percent for loading due to earthquakes in seismic zones 3 and 4 or performance categories C, D, and E.

Framing Stairwells

Stairwells usually are 9 feet 6 inches long or longer and the bay spacing generally is 8 feet. Therefore, all second-floor stairwells must be installed parallel to the frames and perpendicular to the floor joists. First-floor-to-basement stairwells, described in Chapter 5, are below the main frames and can be installed parallel to the floor joists.

Perpendicular-to-joist installations, which are required for the second floor, consist of two 10-inch tie-beams that are 8 feet long, two 10-inch floor beams that are 15 feet long, 8-inch rim track, and the necessary backing plates, bolts, and fasteners. The installation sequence for the second-floor stairwell is basically the same as that for the basement-to-first-floor unit.

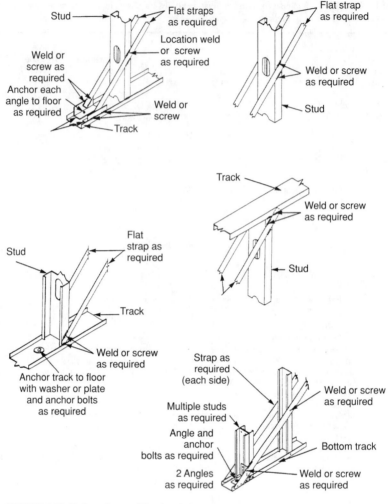

FIGURE 6.45 Various shear-wall bracing designs.

Secure the floor joists to the C-channel beams using #12, ¾-inch self-drilling, self-tapping screws. Floor joists secured to steel wide-flange girders require an 8-inch inverted track fastened with #12, 24-×-1¼-inch screws staggered on 16-inch centers. Fasten the floor joists to the inverted track using #12, ¾-inch, self-tapping screws.

Fasten ¾-inch, tongue-and-groove floor decking to the floor joists using #8, 18-×-1¹⁵⁄₁₆-inch, flathead self-tapping screws on 12-inch centers. It is recommended that the decking be glued to the floor joists with a good adhesive. Also glue the tongue-and-groove joints. Leave approximately ⅛ inch between the endcuts.

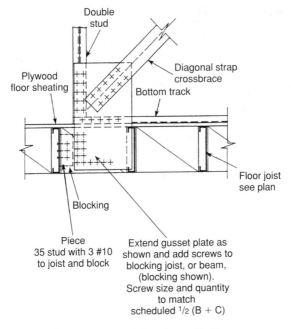

FIGURE 6.46 Stacked second-floor shear wall with common gusset plate.

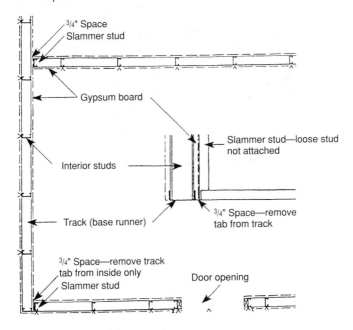

FIGURE 6.47 Typical slammer stud.

FRAMING INTERIOR STUDS

Interior framing follows the same pattern as exterior framing. A clean, clear floor is needed before you can lay out the interior walls. Locate the position of the specific wall section you wish to lay out. Mark the beginning and end points of one side of the wall. Snap a line between these points. Repeat for the other side. Most interior walls utilize 3⅝-inch metal studs, so space the lines accordingly.

Continue this layout procedure for the other interior walls. Mark all locations for passage and closet doors or other wall penetrations. Cut the base track to fit the layout. Use the slammer stud to save material since the interior walls are nonload-bearing (Figure 6.47). Allow ¾-inch spacing on walls where a slammer stud is to be used so that the gypsum board can be installed continuously on exterior and interior walls as applicable.

Position the top track on the ceiling furring directly above the base track (Figure 6.48). For most layouts, you can use a level with the proper length stud to position the top track. For sloped or cathedral-ceiling walls, use a plumb bob to position the top track. Screw the

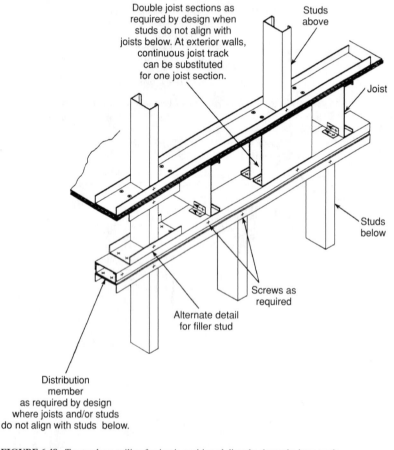

FIGURE 6.48   Top track on ceiling furring is positioned directly above the base track.

top track to the ceiling furring using #8, ½-inch self-drilling panhead screws. Lay out the studs on 16- or 24-inch centers, depending on the local building requirements.

If you are going to wrap the doors with wood, position the metal stud 1½ inches from the edge of the rough openings. Install the headers and cripples as described earlier. Allow for the dimensions of the wood. Install wooden blocking as required on closets, baths, etc., so that heavier items such as fixtures and cabinets can be hung (see Chapter 8).

Attaching Wall Furring Channels

Wall furring channels are used to attach wallboard to masonry walls, steel, or wooden studs. It is used on the exterior of buildings for wall and roof retrofit jobs.

To make a direct channel application, attach the furring channels either vertically or horizontally, on 24-inch centers, to the interior of the masonry or concrete surfaces with hammer-set or powder-driven fasteners, or with concrete stub nails staggered on 24-inch centers on opposite flanges. When the furring channel is installed directly on an exterior wall, and the possibility exists that water might penetrate through the wall, install an asphalt-felt protective strip between the furring channel and wall.

Attaching Gypsum Panels

Attach the gypsum panels either parallel or perpendicular to the channels. Position all the edges over the furring channels in the parallel application; place all ends over the framing in a perpendicular application. Stagger the joints in successive courses. Use the maximum practical lengths to minimize the end joints. Fit the ends and edges closely, but do not force them together. Fasten the panels to the channels with 1-inch type-S screws spaced on 16-inch centers.

Attach the gypsum panels parallel to the channels and so that the vertical joints occur over the channels. Do not use end joints in a single-layer application. Attach the gypsum panels with 1-inch, type-S screws spaced on 16-inch centers in the panel field and at the edges, and with 1¼-inch, type-S screws spaced on 12-inch centers at the exterior corners. For double-layer applications, apply the base layer parallel to the channels and the face layer either perpendicular to or parallel with the channels with the vertical joints offset at least one channel. Attach the base layer with screws on 24-inch centers and the face layer with 1⅝-inch screws on 16-inch centers.

Attach furring strips to the ceiling beam, or in the case of a vaulted ceiling, to the underside of the rafter. Install with two #12, 14-x-¾-inch hexhead, self-drilling screws on 16-inch centers at each attachment point. Frequently, at the endwall, the top track of the exterior wall studs fastens to the furring strips. Because of this, install the furring strips at the end bays before installing the exterior wall studs.

ERECTING CURTAIN WALLS

Nonload-bearing, steel-framed exterior curtain-wall systems have been used for more than 20 years and adapt easily to basic design concepts using conventional materials, methods, and equipment (Figure 6.49). These systems have been specified in all parts of the world for office buildings, schools, shopping centers, motels, hotels, apartments, and some residences.

Steel studs, which are roll-formed from steel and have a corrosion-resistant coating, provide wall-framing members to which interior drywall, veneer plaster bases, and con-

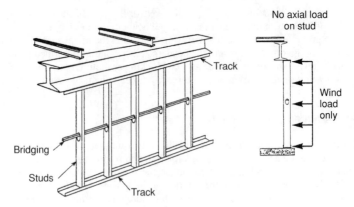

FIGURE 6.49　Typical curtain wall.

ventional plaster systems can be attached (Figures 6.50 and 6.51). They are an ideal backup for stucco and synthetic stucco assemblies because they reduce the dead load by 25 percent when compared to concrete-block backings. The wide choice of stud sizes and spacings accommodates insulation requirements, allows wall heights up to 30 feet, protects against wind loads up to 40 pounds per square foot (psf), and offers a variety of building modules. Requirements for greater wall heights and wind loads usually can be met. Consult the architect and/or a structural engineer for details.

Anchor the studs to the specially designed runners with screws or welds at the top and bottom. Then insert the insulation in the cavity. A special cold-rolled, deep-leg runner, when left unattached to the studs, accommodates floor-slab deflection without imposing an axial load on the studs. Brace studs laterally by screw-attaching gypsum sheathing on the exterior and gypsum panels or base on the interior. Alternately, studs can be braced with straps or cold-rolled channels (CRC) designed according to American Iron and Steel Institute (AISI) specifications.

Exterior surfaces can be unit masonry, portland cement-lime stucco in various decorative panels, or siding materials. Lay brick or other masonry units with a portland-cement mortar and mason's lime and secure every 2⅔ square feet with brick anchors that are screw-attached through the sheathing to the steel studs. This system offers speedier building with the protection of a double-cavity wall and a variety of insulation options.

Figure 6.52 shows several infill curtain-wall application designs. The arrangement of the curtain-kicker details can be seen in Figure 6.53. While light-steel curtain-wall studs are designed for exterior, nonload-bearing curtain-wall systems, they also can be used for interior partitions to provide more rigidity or greater heights than those that can be obtained with standard studs. Height limits for interior partitions using curtain-wall studs are shown in Table 6.4. Attach gypsum wallboard to the full height on both sides of the studs with type-S-12 buglehead drywall screws.

Lightweight, prefabricated, glass-fiber-reinforced concrete panels with 5-inch-thick insulation in the stud cavity offers a 2 fire-resistance rating in addition to high thermal performance.

Ceramic tile, thin brick, aggregated or porcelain-enameled panels, prefinished siding and metal panels, aluminum-framed glass, and exterior insulation systems can be used over this framing. These surfaces and other dry exterior facings that weigh up to 8 psi are applied over sheathing and screw-attached to the studs.

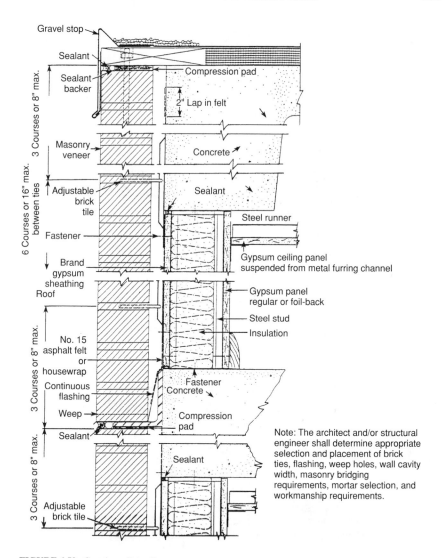

FIGURE 6.50 Curtain-wall details.

ATTACHING SPANDREL WALLS

The spandrel wall is an exterior wall panel, usually placed between columns, that extends from a window opening on one floor to the opening on the next floor (Figure 6.54). Because spandrel walls must be rigidly attached to the primary frame, the designer must allow for vertical movements in the construction of the window header.

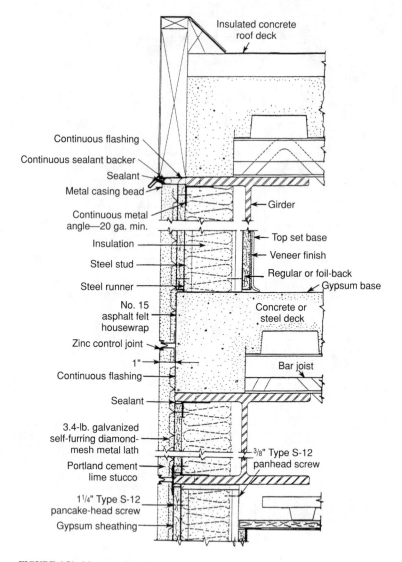

FIGURE 6.51 More curtain wall details.

HANGING WALL SHEATHING

Wall sheathing can be installed when a steel-framed wall is on the ground or in a vertical position. In the normal sequence of construction, wall sheathing is not installed until the roof is framed, sheathed, and the finished roofing material is applied. There are several good reasons for this delay.

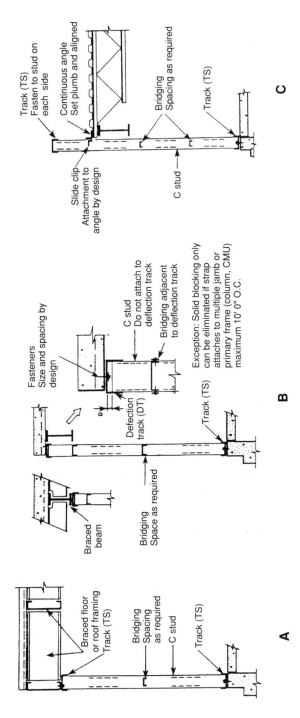

FIGURE 6.52 Infill curtain-wall applications: (A) infill wall; (B) deflection track at infill wall; (C) slide clip at primary frame bypass.

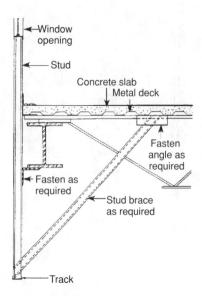

FIGURE 6.53 Curtain kicker details.

- It is extremely important to get the structure under cover so that rainfall does not warp the subflooring and fill the basement or crawl space with water.
- The location of the top edge of the wall sheathing varies with the type of roof and the treatment at overhangs (see Chapter 7).

NOTE: The subject of wall sheathing is covered here only because sheathing is a structural part of the wall and logically belongs in this chapter.
Walls can be sheathed in one of several ways.

- Boards can be applied horizontally or diagonally.
- Sheets of plywood, particleboard, flakeboard, pressed board, oriented-strand board (OSB), or other composition board can be attached.
- Gypsum board can be used.

Plywood is the most common sheathing material. It is durable and has great strength, particularly when applied horizontally. Although plywood comes only in 4-foot widths, it is available in 8-, 9-, 10-, and 12-foot lengths, and in thicknesses from $\frac{5}{16}$ to $\frac{3}{4}$ inch. Most codes permit the use of $\frac{5}{16}$- and $\frac{3}{8}$-inch plywood for sheathing only when it is applied horizontally. The standard material is $\frac{1}{2}$-inch CDX plywood, which is made with exterior glue that can stand quite a bit of moisture. CDX plywood also can be applied vertically.
Gypsum sheathing has three advantages. It does not burn, it does not absorb moisture, and it is low in cost. Its main disadvantages are its weight and limited insulating value. Gypsum sheathing comes in three sizes. Sheets that are 2×8 feet have V-shaped edges that lock together and must be applied horizontally. Sheets that are 4×8 feet and 4×9 feet have square edges and should be applied vertically. Gypsum sheathing is not as strong as plywood and requires more nailing if the corners of the wall are not diagonally braced or sheathed with plywood.

TABLE 6.4 Limiting Heights/Curtain-Wall Studs by Design for Steel Thickness 0.0346 Inch (Minimum Base Steel Thickness 0.0329 Inch)

Stud width	Gauge	Stud spacing	Allow deflection	Partition one layer
2½"	20	16"	L/120	16'2"
			L/240	12'10"
			L/360	11'3"
		24"	L/120	14'8"
			L/240	11'8"
			L/360	10'2"
3⅝"	20	16"	L/120	20'10"
			L/240	16'6"
			L/360	14'5"
		24"	L/120	18'11"
			L/240	15'0"
			L/360	13'1"
4"	20	16"	L/120	21'5"
			L/240	17'0"
			L/360	14'10"
		24"	L/120	19'6"
			L/240	15'5"
			L/360	13'6"
6"	20	16"	L/120	27'4"
			L/240	21'8"
			L/360	18'11"
		24"	L/120	24'10"
			L/240	19'8"
			L/360	17'2"

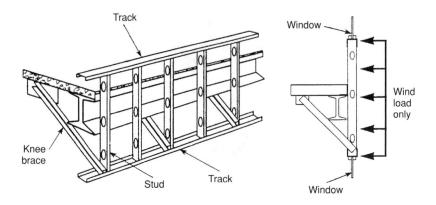

FIGURE 6.54 Typical spandrel wall.

Composition sheathing also is called insulating sheathing because it has a greater insulating value than other sheet materials. The sheets are lightweight, easy to handle, saw, and apply. The material has adequate strength, but is weak in nail-holding power. For that reason, insulating sheathing most often is used on brick-veneer houses. Sheets range in thickness from ½ to 1 inch for wall applications, in width from 2 to 4 feet, and in length from 6 to 12 feet. The most common sizes are 4 × 8 feet with square edges for vertical use, and 2 × 8 feet with interlocking edges for horizontal use.

Apply plywood with screws spaced on 12-inch centers or with adhesive and screws at all stud, track, and column locations. Screw-attach gypsum sheathing to the exterior of each stud with 1-inch, drywall, self-drilling, corrosion-resistant screws spaced ⅜ inch from all ends and edges and on approximately 8-inch centers. Apply a sealant around the sheathing perimeter where it interfaces with other materials and install flashing (Figure 6.55). Install asphalt felt horizontally with a 2-inch overlap and 6-inch endlap and fasten with corrosion-resistant staples. When a stucco exterior is to be applied, take the sheathing in place with screws since the application of the self-furring metal lath completes the sheathing anchorage. Immediately cover all sheathing tacked in this manner with the metal lath.

Sheathing can be accomplished with either of two procedures.

• Precut all openings as the sheathing is installed.
• Install full 4-×-8-foot sheets and cut the opening later.

NOTE: If you cut rough openings after the sheathing is completed, use a router with a plunge-panel bit. Locate the rough opening and plunge the router bit into it close to the bottom or side. Proceed to the bottom or side and pilot the router to cut around the opening.

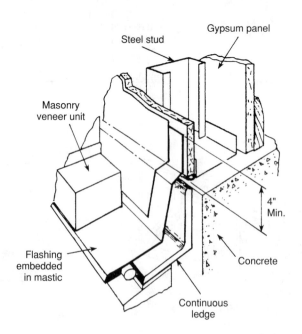

FIGURE 6.55 Method of flashing base.

INSTALLING CEILING JOISTS

The main difference between ceiling joists and floor joists is the loads they support. These loads determine the joist size. Ceiling joists are lighter than floor joists, even when attic space is floored and used for storage. When a roof is carried on trusses, the bottom chords of the trusses replace the ceiling joists. Trusses are even lighter members. The top horizontal member of walls and partitions is usually a doubled top plate. Ceiling joists usually bear directly over studs, although this is not necessary as long as the top plate is doubled. To framers who frame walls, this top wall member is called a *cap plate*.

To framers who frame roofs, however, the top wall member is a *rafter plate*. Roof rafters and ceiling joists rest side by side on the rafter plate. Because rafters are almost always smaller than ceiling joists, the framer must trim the top corners of the joists to the slope of the rafters. Some builders prefer to trim before the ceiling joists are installed. Others prefer to trim after both ceiling joists and rafters are in place. It is standard practice to leave out the end ceiling joists until the roof framing begins because the exact location of these joists varies with the type of roof and the amount of side overhang.

RAISING THE FRAME

Follow several precautions before raising a framed wall that was built on the subflooring. First, devise some means to prevent the wall from sliding off the floor when it is raised. Second, have enough personnel on hand so that the wall can be raised at all points at the same time and at the same speed (Figure 6.56). Third, have braces available to hold the wall upright once it is raised into position (Figure 6.57).

The size of the crew needed to raise a wall varies with the length of the wall. As a general rule, it takes one worker for every 8 linear feet of wall. Put one worker at each end and

FIGURE 6.56 Raising the frame.

FIGURE 6.57   Bracing the wall.

evenly space the others in between. Another worker is needed to nail the braces. If the wall is more than 40 feet long, follow these alternate procedures.

• Build a long wall in sections and raise one section at a time. With this method, the top and bottom plates are not continuous, but are spliced between studs after the sections are in position. Splices should fall at the same point to simplify the construction of the wall.

• Omit window and door frames until the wall is in its final position, plumbed, and braced.

• Use straps that are adjusted by length to keep the frame in balance.

Regardless of the technique used, manually lift the rafter tail and columns to keep the frames straight in the first 15 degrees or so of lift. Place screws in the rafters where you want the straps to help prevent possible movement.

Exterior walls that have been raised into position are far from complete. Securely connect walls that meet at a corner. Straighten all of the walls over their entire length. Complete the nailing. Add cap plates, also called *doublers*, to the top plates to support joists and rafters. Brace corners to withstand winds.

For one-story homes, crossbrace the columns on each sidewall to construct the first bracing bay. Keep the lowest end of each brace above the slab or first-floor sheathing height and connect to the next bay just below the rafter connection.

On two-story homes, connect the lower ends of the crossbraces approximately 2 feet above the first-floor level and then connect them to the next bay 2 to 3 feet above the second-floor support beam. Some one- and two-story homes with center columns require crossbracing in the center also, particularly if they are setting on a central I-beam first-floor system. For homes with clear-span flat ceilings, square the stub columns on the truss structure.

In the procedures in this chapter, exterior wall frames are assembled and erected without cap plates and without sheathing. Many steel-framing contractors do otherwise, particularly volume builders who can temporarily increase or decrease the size of the crew working at a particular site. The exact order of the construction steps is not as important as the result: sturdy walls that meet all code requirements and are both straight and plumb.

FINISHING EXTERIOR WALLS

Once the building is completed to the point shown in Figure 6.58, apply the final exterior materials to the steel studs (Figure 6.59). While the steel-framing contractor usually does not install these finishing products, he or she is interested in what is behind it and how it ties into the steel framing.

Screw-attached sheathing is the usual base for most exterior finishing materials. Fasten the sheathing to the exterior of each stud with 1-inch, self-drilling, corrosion-resistant screws spaced ⅜ inch from the ends and edges and on approximately 8-inch centers. Apply sealant around the perimeter of the sheathing and where it interfaces with other materials. Install flashing as indicated on the plans and specifications. Install asphalt felt horizontally with a 2-inch overlap and a 6-inch endlap and fasten with corrosion-resistant staples as specified. When a stucco exterior is applied, tack the sheathing in place with screws since the applica-

FIGURE 6.58   When framing is at this point, finished exterior can be applied.

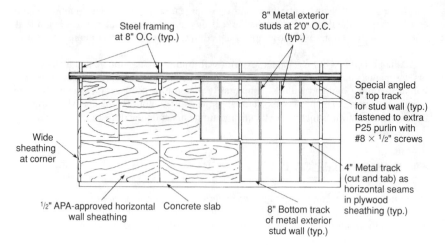

FIGURE 6.59 Applying finished exterior material.

tion of the self-furring metal lath completes the sheathing anchorage. Immediately cover all sheathing tacked in this manner with metal lath.

Apply the metal lath with the dimensions across the supports (Figure 6.60). Lap the ends 1 inch, stagger in adjacent courses, lap the sides ½ inch, and place the lath over the control-joint flange. Screw-attach the self-furring metal through the sheathing and felt to the steel studs spaced on maximum 24-inch centers and runners with 1¼-inch modified-truss or pancakehead, corrosion-resistant screws on 8-inch centers. Screw-attach the lath through the control-joint flanges and sheathing to the studs.

Apply the scratch coat and brown coats to the nominal 1-inch thickness. Apply the scratch coat with sufficient material and pressure to form a full keys-on-metal lath and then crossrake. After the scratch coat sets firm and hard, apply the brown coat to full ground and straighten to a true surface. Leave rough for a finishing coat or matrix application, or seed the aggregate before the brown coat sets.

When masonry materials are installed (Figure 6.61), follow these procedures.

- Erect the masonry materials in a professional manner per the architect's specifications and details.

- Anchor the brick with approved brick ties and screw-attach to each steel stud using two 1¼-inch pancakehead, corrosion-resistant screws. Anchor other masonry units to each stud in a similar manner on maximum 16-inch centers or as recommended by the Brick Institute of America (BIA). If corrugated brick ties are selected, anchor within ½ inch of the bend.

- Support the bricks with steel angles or concrete ledge at each floor or as approved by the architect.

To install panel-type materials, proceed as follows:

- Position the J runners at the floor and ceiling with the short leg toward the inside of the wall. Secure with powder-driven fasteners spaced on 8-inch centers.

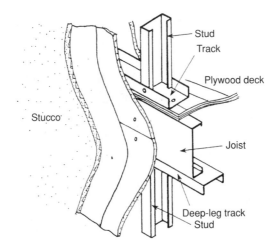

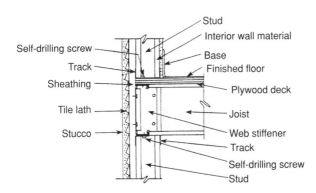

FIGURE 6.60 Application of sheathing to exterior wall.

- Cut the infill panels 1 inch shorter than the floor-to-ceiling height. Drive two 8d duplex nails into the bottom of the gypsum liner panel 4 inches from each edge to support the panel above the runner.
- Install the first gypsum liner panel, position the C-H stud on the free end of the panel, and then continue to alternate the gypsum liner panel and stud applications to complete the wall.

Figure 6.62 details a typical exterior-insulation-finish system (EIFS) on an exterior wall. For other dry-exterior facings, follow the manufacturer's application recommendations.

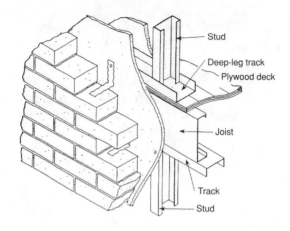

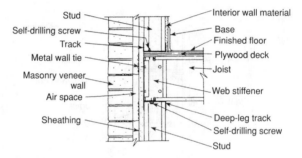

FIGURE 6.61 Installation of masonry materials.

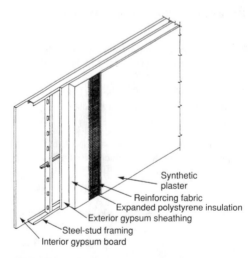

FIGURE 6.62 Typical EIFS exterior wall.

CHAPTER 7
STEEL-FRAMED ROOFS

The architect of steel-framed roofs can use six common roof designs (Figure 7.1) to add beauty and charm to a new residential home: flat, shed, gable, hip, gambrel, and mansard. The roofs vary from a simple flat roof to an intersecting roof with unequal pitches, which is perhaps the most complicated to build. Often a combination of several styles is incorporated into one roof design (Figure 7.2).

With steel framing, these six roof styles are constructed as follows:

Flat Roof. In this type of construction, the ceiling joists support the roof-covering materials. To shed water, the roof is usually made with a slight slope. Since the frame of a flat roof carries the ceiling and roof loads, construction must be sturdy.

Shed Roof. The shed roof is similar to the flat roof. It is pitched in one plane and is often called a lean-to. It can be freestanding or it can lean against another part of a structure, as in the case of a porch.

Gable Roof. The gable roof consists of two straight slopes that rise and meet at the ridge. The triangular wall formed at each end is called a gable.

Hip Roof. This roof style consists of four sloping sides that rise to meet at the ridge. If the structure is square, all four sides meet at the ridge to form a pyramid roof.

Gambrel Roof. The gambrel roof is a variation of the gable roof. It has two slopes on each side, but the lower slope is steeper than the upper.

Mansard Roof. The mansard roof is a variation of the hip roof. It has two slopes on each of the four sides. As with the gambrel roof, the lower slope is steeper than the upper.

Whatever the style of the roof, certain framing terms are used to express the shape of the roof and to establish the length of its parts. These terms are *span, run, line, rise, pitch,* and *pitch line* (Figure 7.3).

The *span* of a roof is the horizontal distance from the outside edge of the rafter plate on one wall to the outside edge of the rafter plate on the opposite wall. The *run* of a roof is the horizontal distance from the outside edge of a rafter plate to the centerline of the ridge. If the slope of the roof is the same on both sides of the ridge, and the eaves are the same height, the run is equal to half the span. The *pitch line* is an imaginary line that runs through the outer edge of the rafter plate to the ridgeboard and parallel to the edges of the rafter.

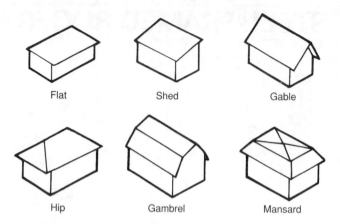

Flat   Shed   Gable

Hip   Gambrel   Mansard

FIGURE 7.1 Common steel-framed roofs.

FIGURE 7.2 Metal-framed roof can have more than one roof design.

The *rise* of the roof is the vertical distance from the top of the rafter plate to the point where the pitch line intersects the centerline of the ridge. The rise, run, and pitch line form a right triangle. The hypotenuse of this triangle is the line length because it makes no allowance for cuts at the ridge and eaves, nor for any overhang.

The *pitch* of a roof is its slope and the ratio of the rise to the run. On drawings, the pitch usually is shown on an inverted triangle as two numbers. The horizontal number is always 12. The roof slope chart in Figure 7.4 converts rise/run into degrees of slope.

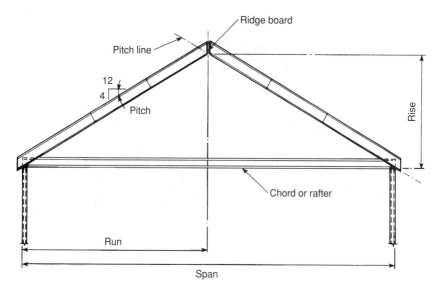

FIGURE 7.3   Common roofing terms.

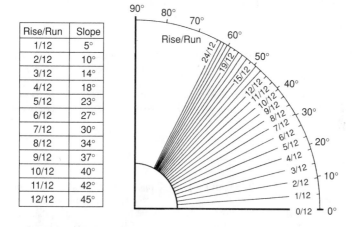

Rise/Run	Slope
1/12	5°
2/12	10°
3/12	14°
4/12	18°
5/12	23°
6/12	27°
7/12	30°
8/12	34°
9/12	37°
10/12	40°
11/12	42°
12/12	45°

FIGURE 7.4   Roof slope chart for converting rise/run into degrees of slope.

DESIGNING THE FRAMING PLAN

Roof framing requires that major load-carrying members, such as ridges, valleys, and hips, be specifically designed to support the roof rafters. All roof designs require a framing plan. Figure 7.5 shows a simple plan. Figure 7.6 shows an actual, slightly more complex, roof assembly elevation that was generated with a computer and then the actual framing.

When framing a roof, the contractor can use either trusses (Figure 7.7) or stick-by-stick construction (Figure 7.8). Roof purlins often are employed with both types of roof framing.

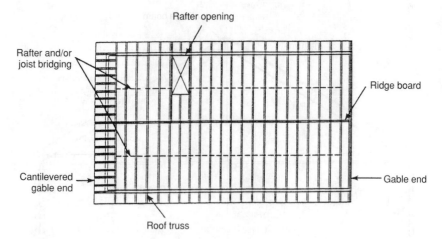

FIGURE 7.5   Simple roof-framing plan.

Let's look at the steps for constructing a stick-for-stick steel-framed gable roof, which is the easiest of the pitched roofs to build. Start by assembling all the steel parts. Mark the rafter locations on the top plate and ridge board (Figure 7.9). Two other connection methods at the ridge are shown in Figure 7.10, while Figure 7.11 illustrates typical eaves details. If the rafter spacing is the same as that for the ceiling joists, install the rafters adjacent to each joist. Screw one end of the rafter to the plate while another worker supports the free end (Figure 7.12). Do the same for the opposite rafter and then fasten the top ends of the rafters at the ridge. On a vaulted or cathedral ceiling, tie the rafters into the ridge board above the balcony (Figure 7.13).

In high-wind areas and hurricane-prone zones, brace the roof frame as shown in Figure 7.14. The braces act as trusses and reduce the span of the rafters. Use collar-beam ties to reinforce the roof frame (Figure 7.15).

Framing the Gable End

Install the gable-end studs flush with the endwall plate. These walls can be framed in three ways (Figure 7.16). If there is no opening in the gable for a window or ventilating louver, place the upper studs directly above the studs in the lower wall. This placement is not a structural requirement, however. It usually is better to begin at the gable's centerline and frame in both directions.

The second method is to center a stud directly under the ridge board. This placement works well if a pair of windows or a standard 30-inch window is centered in the gable.

The third method is to place studs 8 inches on each side of the centerline. This placement works well if a standard 46-inch window is centered in the gable. Because the studs that are equidistant from the gable's centerline are the same length and are cut the same way, framers can save time by planning their work from a centerline.

All the studs in half a gable are a different length. But the difference in the length of adjacent studs is the same. This constant is known as the *common difference*. To find the common difference between studs in inches, multiply the unit rise of the roof by the stud spacing stated in feet. Suppose that a roof has a 4/12 pitch and that the studs are on 16-inch

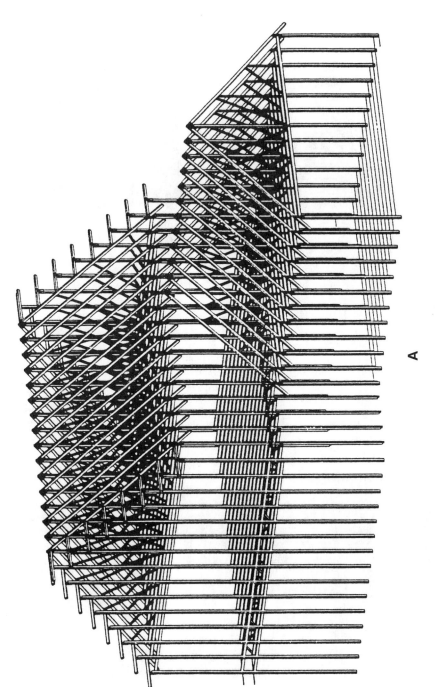

A

FIGURE 7.6 (A) Computer-generated complex roof framing plan, and (B) plan as constructed.

B

FIGURE 7.6 *(Continued)*

FIGURE 7.7 Use of truss.

FIGURE 7.8 Workers installing a stick-by-stick roof.

FIGURE 7.9 Rafters installed between plate and ridge board.

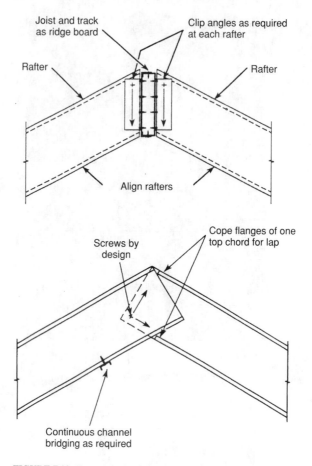

FIGURE 7.10   Two methods of connecting rafters at ridge.

centers in the gable. If unit rise × stud spacing in feet = the common difference in inches, then 4 × 1⅓ = 5⅓.

Align the studs with those in the wall below. If a window is to be installed in the gable end, measure one-half the rough opening width at each side of the gable center. Start the first studs outside these marks. Space the balance of the studs at the regular interval. Mark the diagonal line by tracing along the underside of the rafter. Be sure to hold the stud plumb when marking.

As already mentioned, the studs of the gable endwall have a common difference as they decrease in length. The difference between the two adjacent studs is the common difference for all the other studs. Lay out the remaining studs by increasing or decreasing each succeeding stud by this amount. Screw the studs flush with the end plate and rafter.

Figure 7.17 illustrates the gable end and eaves portion of a cathedral ceiling. Figure 7.18 shows typical cantilevered gable ends. When the roof plans call for a gable overhang, or *extended rake*, the end rafter is supported by short lookouts. These are constructed as shown in Figure 7.19.

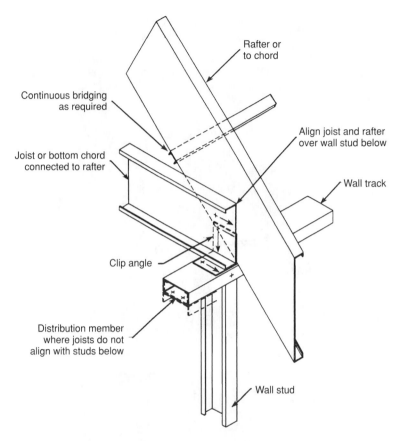

Rafter or
to chord

Continuous bridging
as required

Align joist and rafter
over wall stud below

Joist or bottom chord
connected to rafter

Wall track

Clip angle

Distribution member
where joists do not
align with studs below

Wall stud

FIGURE 7.11 Roof-eaves details.

Large roof openings require the interruption of the roof framing and are treated the same as floor openings (see Chapter 5). Place headers between the rafters with their faces plumb. For small openings, install the rafters in the usual manner. Cut the openings and frame afterwards. Construct large openings during the framing, but do not alter the rafter spacing. Double the headers and add trimmers to each side of the opening.

For chimney openings, line up the headers in the roof with those in the floor below. Plumb the lines carefully and then install the headers. Allow a 2-inch clearance all around so the framing clears the masonry.

Squaring the House

After all the frames are standing and the supports or purlins are installed to stabilize the frames, it is time to square the house (Figure 7.20). To do this, install a ceiling furring strip along the bottom of the ceiling joists. Place marks on the furring strips to assure proper spacing. Make sure you install the furring in the location needed for the ceiling construction. If you marked the lines on the ceiling beams when they were on the

FIGURE 7.12   Roof work is usually a two-person operation.

FIGURE 7.13   Installed rafters in a vaulted or cathedral ceiling.

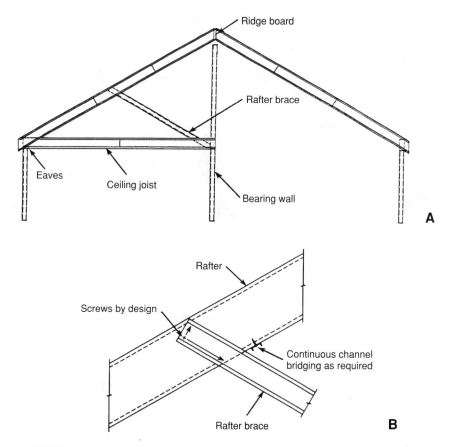

FIGURE 7.14 (A) Typical rafter braced roof assembly, and (B) method of fastening the brace to the rafter.

ground, you have already determined the locations. Position the furring strip on the line closest to the center.

Install the crossbracing at the peak between the stub columns of the brace frame. Two 10-foot purlins are included for this purpose. In a straight home, as opposed to an L-shaped home, you need only do this once. The balance of the frames should straighten right up with the proper spacing of the purlins and ceiling furring.

Check columns for plumb in both the length and width of the house. Use come-alongs, or cables and turnbuckles, to straighten any frame that is not square. After the house is fully square, install the second purlin screw in each purlin, and then install the rest of the purlins. Do not remove the come-alongs used to square the frame until the house is sheathed.

Installing the Purlin End Caps

Install the purlin end cap after the house has been squared. Because the ends of the purlins might not align exactly, snap a chalkline from the peak to the eaves 1½ inches in

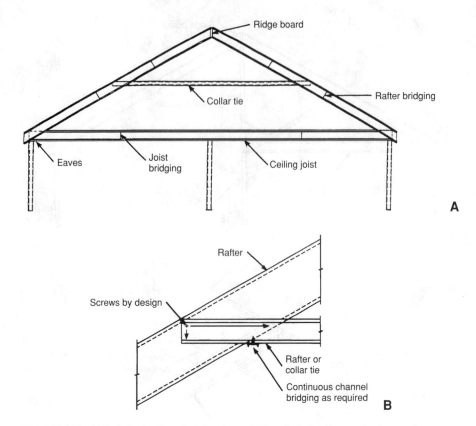

FIGURE 7.15 (A) Typical rafter-framed roof section, and (B) method of making a collar tie to a rafter.

from the longest purlin ends. Install the leading edge of the end cap along this line with one #8, 18-×-½-inch panhead, self-drilling screw at each purlin (Figure 7.21).

The eaves cap can be installed two ways, as either a square fascia (Figure 7.22) or a plumb fascia (Figure 7.23). The square fascia is square to the rafter tails and produces an angled appearance. The plumb fascia is parallel to the sidewalls and perpendicular to the ground. The plumb fascia facilitates the installation of gutters, however the dimension from the fascia line to the steel line is 2 feet 2 inches. Depending on the siding used, you cannot get two pieces of soffit from a 4-×-8-foot sheet of soffit material. The eaves cap is installed with #8, 18-×-½-inch panhead, self-drilling screws on 2-foot centers.

Attach the finished wooden fascia directly to the eaves cap. Because the fascia reflects any waviness in the eaves cap, install the eaves cap with a stringline. Run the stringline from one endwall purlin end cap to the other. Space the stringline ½ inch from the edge of the first purlin.

For the square-cut fascia, place a framing or speed square against the angle cut of the rafter tail and slide the purlin against the square. Clamp in place and fasten with the recommended screws.

For the plumb-cut fascia, place a framing square at the base of the rafter and slide the purlin against the square. Clamp in place and fasten with the recommended screws.

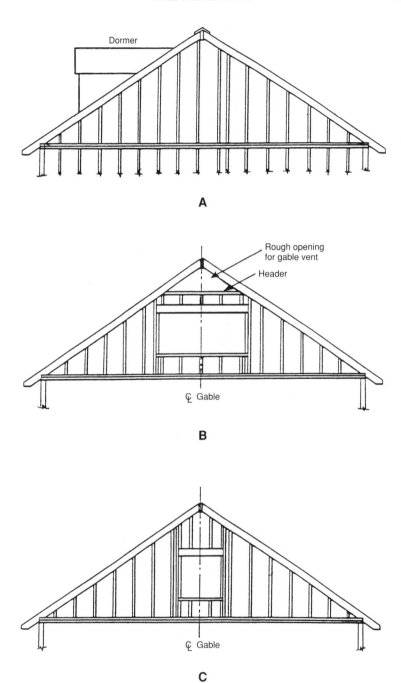

FIGURE 7.16 Three methods of framing a gable end: (A) stud above stud; (B) with centerline of gable on the centerline of the middle stud; or (C) with the centerline of the gable end between a pair of studs.

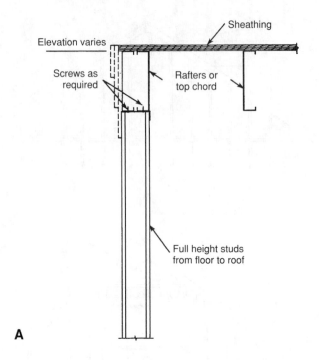

A

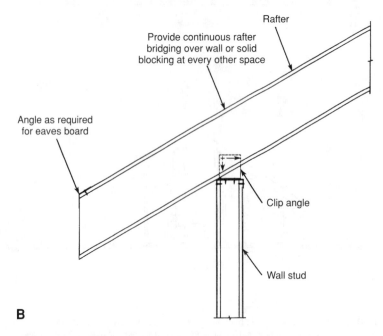

B

FIGURE 7.17 The (A) gable end and (B) eaves portion of a cathedral ceiling.

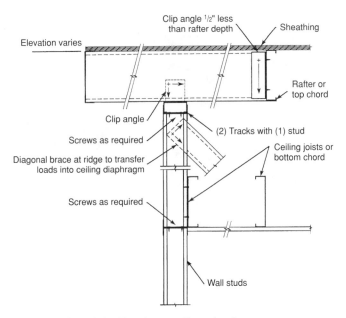

FIGURE 7.18 Typical gable end on a cantilevered roof.

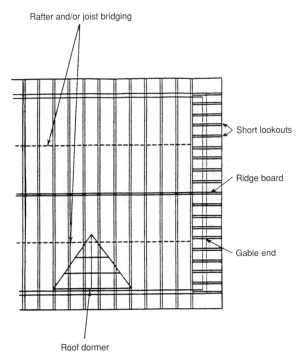

FIGURE 7.19 Support end rafters with short lookouts when the roof plan calls for a gable overhang.

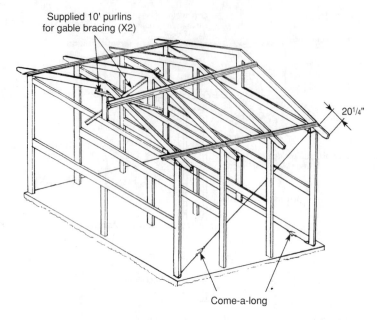

FIGURE 7.20   Squaring the house.

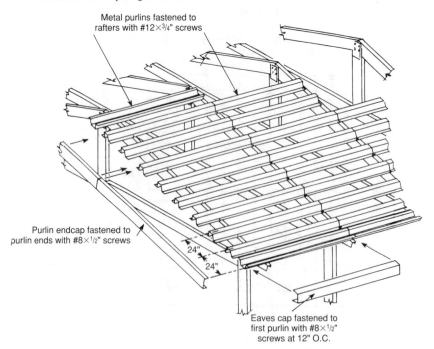

FIGURE 7.21   Installing the purlin end caps.

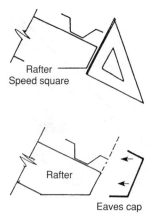

FIGURE 7.22 Installing eaves caps as a square fascia.

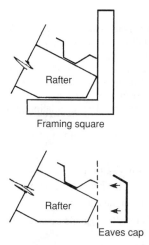

FIGURE 7.23 Installing eaves caps as a plumb fascia.

NOTE: Preassemble the eaves purlin and eaves cap on the ground and install it as a unit. Assure straightness by using a stringline to align the outer top edge of the eaves-cap assembly.

Constructing a Soffit

The soffit-framing members are composed of 3⅝-inch metal stud and 2-×-2 light-gauge soffit angle (Figure 7.24). The steel stud is furnished in stock 10-foot lengths for field cut-

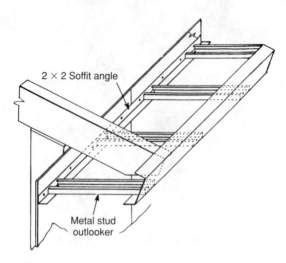

FIGURE 7.24 Metal stud and 2-×-2 light-gauge soffit angle.

ting to nominal 2-foot outlookers and the 2-×-2 angle is furnished in either 10- or 12-foot lengths. Take care when installing the soffit angle to be sure it is level and on the same plane as the bottom of the eaves cap.

Fasten the 2-×-2 soffit angle through the wall sheathing to the metal studs with one #6, 20-×-1¼-inch ply-metal self-drilling screw on 2-foot centers. Then install the outlooker between the eaves cap and the soffit angle on 2-foot centers with one #8, 18-×-½-inch pan-head, self-drilling screw at each end. The installation procedure is the same for both the eaves and gable soffit.

To construct the returns, or *bird boxes*, at the corner of the gable roof, proceed as follows (Figure 7.25).

- Extend the 2-×-2 angle past the endwall a distance equal to the overhang of the purlin and eaves cap. This establishes the base of the return.
- Tab a piece of 2-×-2 angle from the extended angle, or plumb, to the purlin eaves cap.
- Install a steel stud outlooker or 2-×-2 angle between the extended 2-×-2 angle and the eaves cap.
- Install a second plumb 2-×-2 angle parallel to the other vertical support and fasten it to the roof sheathing.
- Install another 2-×-2 angle horizontal to and just below the height of the eaves outlooker to complete framing the box.

Steel-framed exterior gypsum panels that are ½ or ⅝ inch thick, also are suitable for commercial use as exterior ceilings for covered walkways and malls, large canopies and parking areas, and for residential use in open porches, breezeways, carports, and other exterior soffit applications. These areas must be horizontal or slope downward away from the building.

When building an exterior soffit, use the longest practical board lengths to minimize the number of butt joints. Framing should be on no more than 16-inch centers for ½-inch-thick gypsum board, and not more than 24 inches on center for ⅝-inch board. Allow a ¹⁄₁₆- to ⅛-inch space between the butted ends of the board. Fasten the panels to the supports

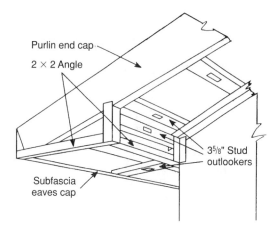

FIGURE 7.25 Birdbox soffit construction.

with screws spaced on 12-inch centers or nails spaced on 8-inch centers. For steel framing, use 1-inch type-S corrosion-resistant screws or type-S-12 for 20-gauge or thicker steel. A closed soffit is detailed in Figure 7.26.

Provide suitable fascias and moldings around the perimeter to protect the board from direct exposure to water. Unless protected by metal or other water stops, place the edges of the board not less than ¼ inch away from the abutting vertical surfaces. At the perimeter and vertical penetrations, cover the exposed core of the panels with metal trim or securely fasten moldings for weather protection (Figure 7.27).

A suspended metal framework for the support of the ceiling board usually is composed of 1½-inch steel channels for the main runners and either steel furring channel or steel studs for crossfurring (Figure 7.28). Space the main runners as follows:

- Maximum 4 feet on center if using furring channels
- Maximum 6 feet on center for 2½-inch steel studs
- Maximum 8 feet on center for 3⅝-inch metal studs

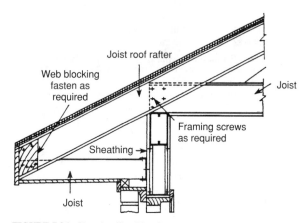

FIGURE 7.26 Closed soffit with attic vent.

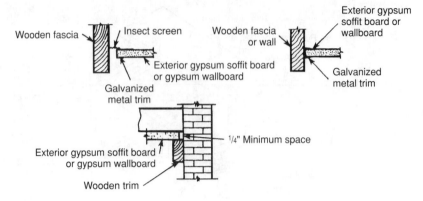

FIGURE 7.27   Provide fascias and moldings to protect from water exposure.

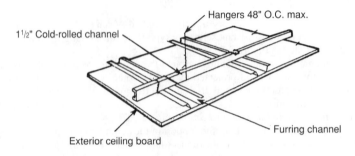

FIGURE 7.28   An exterior ceiling-board suspended soffit.

Suspend the main runners with a minimum 8-gauge galvanized wire spaced on maximum 4-foot centers. Secure the metal furring channels to the main runners with 1½-inch channel clips. Alternate the sides at each intersection or saddle-tie with double strands of 16-gauge galvanized tie-wire. Secure the metal stud furring to the main runners with double-strand 16-gauge tie-wire on maximum 16-inch centers (Figure 7.29). For added rigidity, nest an 8-inch-long piece of track at each tie-wire location.

The main runners and furring must have a minimum clearance between their ends and any abutting structural element. Locate a main runner within 6 inches of the parallel walls. Support the ends of these runners with hangers spaced not more than 6 inches from the ends. The closest edge of the furring member must be no more than 2 inches from the parallel wall. At any openings for vents, light fixtures, etc., that interrupt the runners or furring, provide reinforcement to maintain support equal to that of the interrupted members. Apply the gypsum-board soffit with 1-inch type-S screws on maximum 12-inch centers.

When the area above the ceiling board opens to an attic space above habitable rooms, vent the space to the outside in accordance with accepted recommendations of 1 square foot of free vent area per 150 square feet of attic area. When the ceiling board is applied directly to the rafters or to the roof ceiling joists that extend beyond the habitable rooms, as in flat-roof construction, vents are required at each end of each rafter or joist space. Screen the vents so that they are a minimum of 2 inches wide by the full length between the rafters or joists. Attach the vents through the board to 1-×-2-inch (minimum) backing

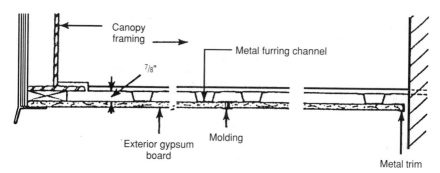

FIGURE 7.29 Typical steel-furred canopy.

strips, which were installed prior to the board application. Frame the vent openings and locate them within 6 inches of the outer edge of the eaves.

As already mentioned, when the ceiling-board expanse exceeds 4 feet, provide a space of at least ¼ inch between the edge of the ceiling board and the adjacent walls, beams, columns, and fascia. This space might be screened or covered with molding, but must not be caulked.

Ceiling gypsum board, like other building materials, is subject to structural movement, expansion, and contraction due to changes in temperature and humidity. Install either a casing bead with a neoprene gasket, a manufactured control joint (Figure 7.30), or a con-

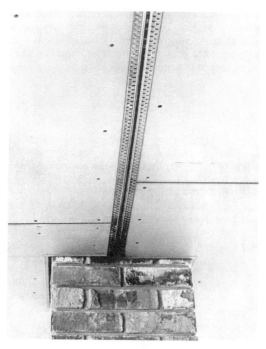

FIGURE 7.30 Typical manufactured control joint.

trol joint that consists of two pieces of metal trim back to back in the ceiling board where the expansion or control joints occur in the exterior wall or roof. Aluminum H moldings can serve as control joints as long as the board is not inserted tightly.

Space control joints no more than 30 feet apart in long, narrow areas. Separate the wings of L-, U-, and T-shaped areas with control joints. These joints usually are placed to intersect light fixtures, vents, etc., and relieve stress concentrations. Design a canopy to resist uplift.

BUILDING HIP ROOFS

The hip roof can be described as a gable roof that has been sliced diagonally from each end in a plane lying downward and outward from the ridge (Figure 7.31). The line formed where the two sloping sides meet is called the *hip*. The corner rafters of such a roof are called *hip rafters*. The hip rafter is shown in plan view in Figure 7.32.

The hip frames, when raised and bolted together, become rigid and stable (Figure 7.33). When properly installed and after the purlins are connected to adjacent frames, the hip frame ties the structure together and simplifies the plumbing and bracing of the other frames. It is necessary to locate the anchor bolt at the correct position and elevation, otherwise plumbing and squaring are difficult. Prior to assembly, mark all hip rafters except the main double-corner rafters with the same purlin locations used on the rest of the house. Also mark the ceiling beams.

To raise the hip frames, lift the frame with the installed hip-connection plate and brace. The hip-connection plate faces the endwall of the hip and connects with the main hip rafters. Do not preassemble the short rafters prior to raising. Raise the next hip frame, including only the columns and ceiling beam. Install the temporary bracing.

Assemble the endwall hip section, including the corner columns, ceiling beams, and center column, and tighten fully. Raise the endwall into position. The endwall can free-stand long enough to proceed with bolting together the rafters. Tighten the anchor bolts.

FIGURE 7.31 Typical hip roof.

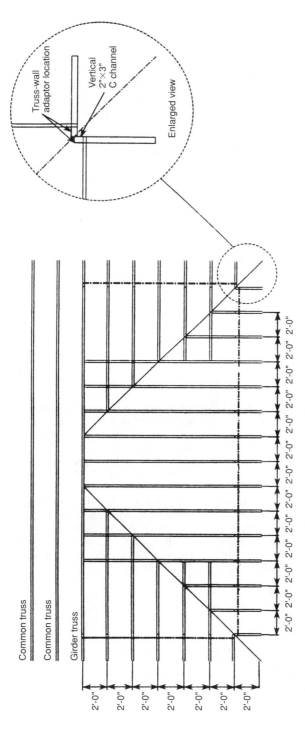

FIGURE 7.32 Plan view of a hip rafter.

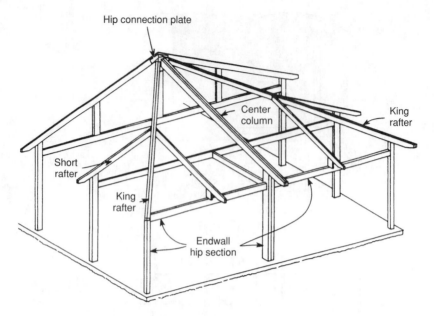

FIGURE 7.33 Typical hip frames.

NOTE: It is not necessary to install any purlins on the hip section before raising the rest of the structure.

Attach a nylon strap to the center common rafter so that the peak balances higher than the rafter end. Raise and bolt the top connection and align with a spud wrench. Then bolt the lower end of the rafter to the double column. You might wish to slightly loosen the center-column anchor bolts to ease installation.

Now use the same procedure to raise the double-corner hip, or *king*, rafters into place. Tighten all bolts. Raise the short tie-rafters with stub columns, as applicable, into position after all the hip pieces are in place and fastened.

To install the purlins on the hip angle, cut them on a 45-degree angle using a 14- or 16-inch chopsaw. To determine the length of the purlin, snap a chalkline out to the corner hip rafter using the marked rafters as a guide. Measure from the purlin on the previous rafter to the location marked on the corner rafter, ⅛ inch short of center. Center it on the previous straight rafters, which usually are the tie-rafters. Use this dimension to start. If the roof pitch is 6/12, the difference in length for the next purlin, as you proceed up the angled rafter, is 21½ inches.

Once the purlin installation is complete, install 2-×-2 angles to complete the framing. Place the 2-×-2 angles on top of the purlin at the 45-degree cut end to provide a screwing surface for the decking that runs from eaves to peak and parallel to the king rafter.

The intersecting hip roofs can be of equal or unequal pitch. The equal pitch exists when the span and slopes of the two roof sections are equal. In such a roof, the common rafters of both sections are equal and the valley rafters butt against both ridges (Figure 7.34). If the span of the intersecting roof is less than that of the main roof, only one valley rafter extends to the ridge of the main roof (Figure 7.35).

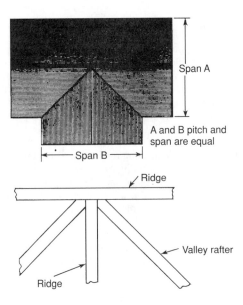

FIGURE 7.34 Equal pitch and open-hip roof.

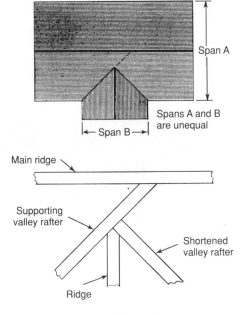

FIGURE 7.35 Intersecting hip roof.

Framing the Saddle Valley

Valley rafters are similar to hip rafters. They are placed where two sloping surfaces of the intersecting roofs meet to form an angle. The saddle is that portion of the roof that connects a surface that slopes to a vertical surface (Figure 7.36). The materials used in this construction are 8-inch studs, track, 2-×-2 angle, and purlins. All these materials must be field cut to meet the particular roof application. Cut the bottom of the studs to the slope of the roof. Miter cut the other materials as required. Before constructing the saddle, completely sheath the receiving roof.

Locate several points to establish chalklines from which to work (Figure 7.37). Begin by extending the ridgeline of the connecting roof to the point at which it meets the receiving roof. Attach a stringline to the connecting roof ridge and extend it to the receiving roof using a line level to maintain a level ridge. Attach the string to the receiving roof and leave it in place. Have a second worker sight down the connecting ridge to assure that the stringline maintains a straight ridge for the saddle. The point at which the string attaches to the receiving roof is the point where the saddle sheathing meets. Measure and note the length of the string back from this point to the connecting frame.

Next, find the center of the connecting frame and mark it on the edge of the receiving roof adjacent to the connecting frame. Snap a chalkline from this point to the string connection point. This is the center of the saddle and should be directly below the stringline.

To establish the valley points, from the connecting frame center mark to each side, mark the receiving roof at the string length dimension noted previously. Chalk a line from each valley point to the string connection point. These are the valley lines where the saddle sheathing meets the receiving roof.

Locate and chalk the valley's short stud walls, or *set* lines, on the receiving roof. Locate them a maximum of 8 feet on center and adjust so that they fall directly over a purlin on the receiving roof (Figure 7.38). Now measure the distance from the stringline down to the intersection of the valley set and chalked ridgeline. Deduct the height of the purlin and

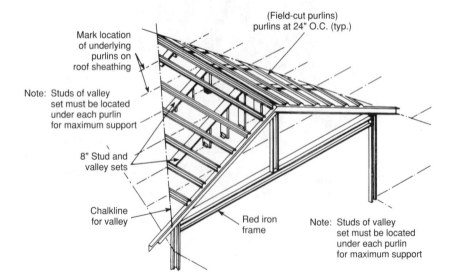

FIGURE 7.36   Valley rafters butt against both ridges.

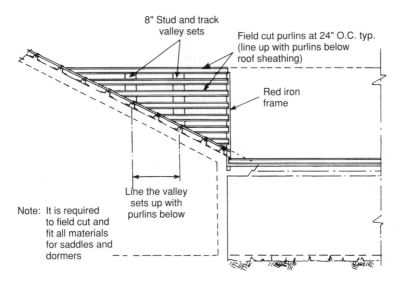

FIGURE 7.37 One rafter extends to the ridge of the main roof.

sheathing from this measurement to produce the center-valley set-stud length. Install this stud on the receiving roof using a 6-inch-long piece of 2-×-2 angle.

Then install the top track of the valley set. This track has to be tabbed to attach to the receiving roof and fall short of the chalked valley line, depending on the slope of the roof

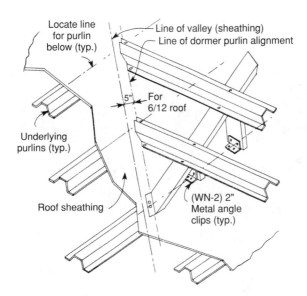

FIGURE 7.38 Saddle valley framing.

and the height of the purlins. After the track is connected to the center stud and the receiving roof deck, install the two purlins closest to the ridge. Make sure they are level with the stringline. Minor adjustments in height might have to be made at this point. Repeat this procedure for each valley set on the receiving roof. Once they are in place and the purlins are level, install the rest of the valley set-studs out in each direction from the center. Space the studs directly below where the purlins bear on the track. After all the studs are in place, complete the purlin and roof sheathing installation.

Building Dormers

A dormer projects above the main roof of a home and can enclose living space, add to attic ventilation, or be strictly decorative. While there are many different configurations, the most common types are the shed and gable. The shed dormer provides greater use of floor space in the upstairs rooms. The width of the shed dormer is not limited, as it can extend almost the full length of the roof (Figure 7.39).

When the shed roof extends to the ridge, make a top cut at the top of each rafter. If the shed roof stops short of the ridge, install a double header across the top of the opening. The front wall usually extends from the main wall plate. Install double-trimmer rafters to support the sidewall studs.

The gable dormer is like a miniature roof and might contain a window or ventilator (Figure 7.40). Similar to the shed dormer, it can extend from the plate, but is usually set back from the plate of the main roof and stops short of the ridge. Construct the double headers and trimmers in the usual manner. Place the headers between the main roof rafters at a point where the dormer ridge and valley rafters meet. The slope of the dormer roof should match that of the main roof.

USING ROOF TRUSSES

Trussing, or triangulated framing, adds stability to a lightweight steel frame. Historically, the truss is used to achieve the simple, double-slope, gabled roof. This typically is done by

FIGURE 7.39 Typical shed dormer.

FIGURE 7.40   Typical gable dormer.

using sloping members and a horizontal bottom member (Figure 7.41). It also is the most common method of roof construction employed by steel framers.

The use of roof trusses has been well established in light commercial and residential construction over many years. Trusses permit wide spans without intermediate supports. This is a great advantage, especially in architectural designs that call for spacious rooms without supporting partitions or posts. Their use also permits rapid assembly of the roof frame, which enables the structure to be enclosed and protected in less time.

The truss or trussed rafter consists of upper and lower chords and diagonals. The upper chords represent the rafters and the lower chords correspond to the ceiling joists. The di-

FIGURE 7.41   The roof trusses are one of the most popular methods for roof construction.

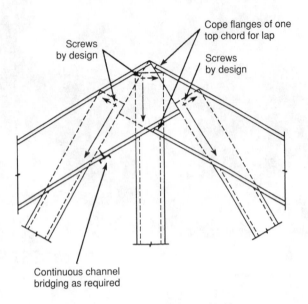

FIGURE 7.42 Two typical web connections to the ridge.

agonals or webs serve as braces. Figure 7.42 shows a typical two-web connection to the ridge at the top chord and the bottom chord of a truss.

Architects and general building contractors have long been aware of the quality and cost-saving benefits of shipping these preassembled building components to the job site ready for erection. Today, however, more architects are turning to lightweight steel roof trusses as a cost-effective, high-quality alternative to their wooden predecessors.

This new interest is due to several market developments, such as the fire-resistant qualities of steel-framing trusses, increased use of metal roofing systems (Figure 7.43), the retrofit of older flat roofs with pitched roofing systems, and augmented awareness of the inherent quality and strength-to-weight ratio of lightweight steel construction (Figure 7.44).

Architects have discovered a series of benefits offered by steel roof trusses. For instance, lightweight steel framing offers precision and product quality, nearly unlimited length capability with roll-formed steel members, and excellent compatibility with most roofing systems. Steel components are being specified for shopping centers, motels, smaller commercial buildings, and multifamily housing, as well as residential homes. These benefits, along with job-engineered economy, offer the designer a unique combination of flexibility, quality, and price. Because of these features, more and more jobs are being designed with steel trusses, which in turn is creating a market opportunity for contractors willing to take on this new challenge in roofing systems.

Trusses can be used in a number of ways as part of a building's total structural system. Figure 7.45 illustrates the elements of a truss. Figure 7.46 shows a series of single-span, planar trusses and the other elements of the building structure that develop the roof system and provide support for the trusses. In this figure, the trusses are spaced a considerable distance apart. In this situation, it is common to use purlins to span between the trusses. Support the top-chord truss joints to avoid bending in the chords. The purlins, in turn, support a series of closely spaced rafters that are parallel to the trusses. Attach the roof sheathing to the rafters so that the roof surface actually floats above the level of the top of the trusses.

FIGURE 7.43 Typical metal roofing application.

FIGURE 7.44 Trusses used in roof retrofit work.

Selecting Truss Types

Truss types, like buildings, can vary greatly. Typically, the building's use determines, to a certain extent, the appearance of the structure, including the shape of the roof.

Trusses are assembled in various designs. A pitched truss is built with web variations (Figure 7.47). They are popular for light commercial and some residential spans up to 80 feet with 15- to 20-foot spacing. Pitched trusses can be built for longer spans, but they are

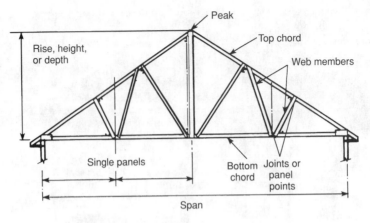

FIGURE 7.45 Elements of truss.

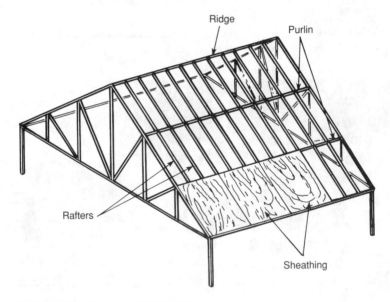

FIGURE 7.46 Single-span, planar trusses.

most economical when compared with other types of trusses in spans to about 60 feet. Generally, the design employs normal slopes or truss depths that are ¼ to ⅛ the span.

Trusses that are commonly used in steel residential framing are the *Finke*, also known as *W type*, *Belgian*, and *Kingpost* (Figure 7.48). Shed trusses are used horizontally or vertically depending on the need (Figure 7.49). Stepped trusses are placed on a hipped roof at the shorter ends (Figure 7.50). Scissors trusses ideally are suited for houses designed to have a clerestory or slanting cathedral-ceiling effect (Figure 7.51). Figure 7.52 shows the placement of the webs at the ridge and at the bottom chord intersection on a scissors truss. Other methods for connecting the truss-bearing surface at the eaves are shown in Figure 7.53.

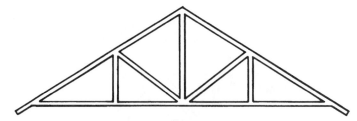

FIGURE 7.47 Pitched-designed truss.

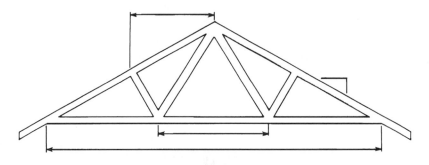

A. Fink truss (W type)

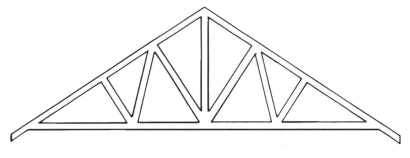

B. Belgian truss

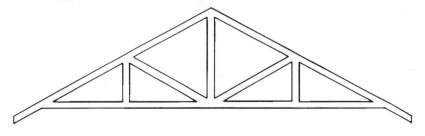

C. Kingpost truss

FIGURE 7.48 Common types of roof trusses used in steel residential framing.

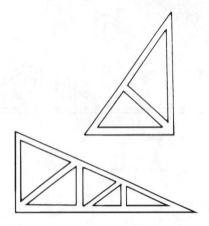

FIGURE 7.49 Typical shed trusses.

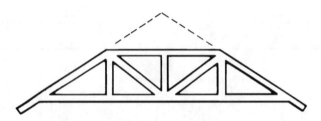

FIGURE 7.50 Typical stepped trusses.

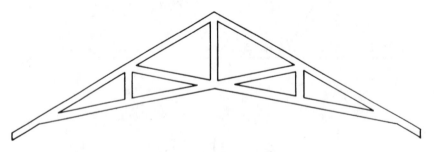

FIGURE 7.51 Typical scissors trusses.

Flat roof systems, such as the *Pratt*, *Howe*, and *Warren* (Figure 7.54), can be used for both floor and roof construction. Use special connector plates to fasten the members to the truss. The trusses can then be ganged, or fastened into groups, for hoisting and securing. When fastening multiple trusses, press the screws into the members to create an extra measure of rigidity. Shopping centers are usually designed with a flat roof, which creates a need for definition in the building's appearance. In this design, shed trusses are attached to the front of the building or set on top of the roof. A dormer, ei-

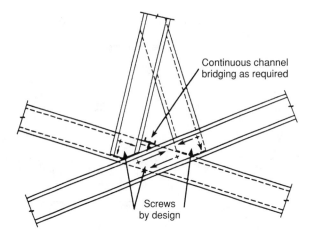

FIGURE 7.52 Placement of webs at ridge and bottom-chord intersection on a scissors truss.

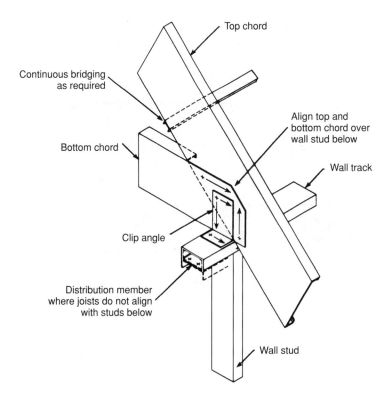

FIGURE 7.53 Methods of connecting truss-bearing surface at the eaves.

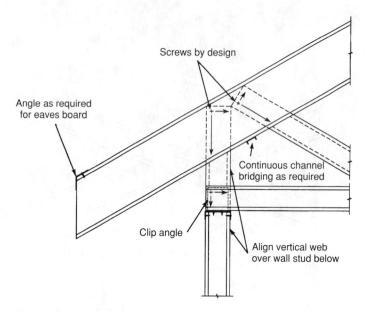

FIGURE 7.53 *(Continued)*

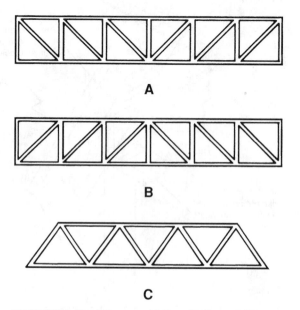

FIGURE 7.54 Flat roof systems: (A) Pratt, (B) Howe, (C) Warren.

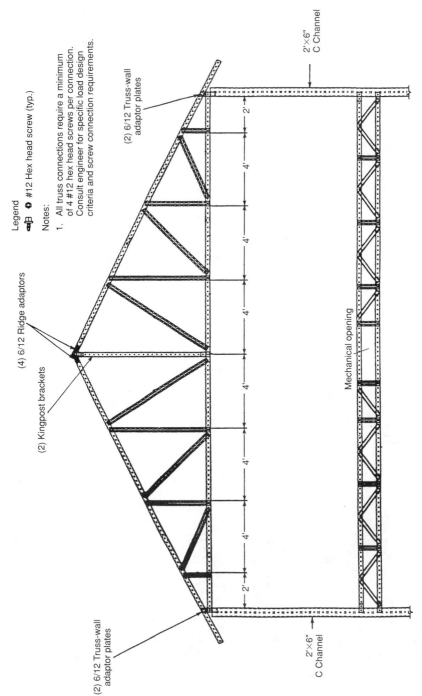

FIGURE 7.55 Using a flat truss as a floor beam.

ther real or false, then is added to the slope to achieve the desired look. As shown in Figure 7.55, a flat truss can be used as floor-beam warping in junction with a sloped roof truss. Another common application suitable to most structures is the pitched roof. A gable roof, which is pitched two ways, or a hip roof, which is pitched four ways, is built with prefabricated trusses set directly atop the bearing walls, sheathed, and then finished with roofing material.

The lightest open-web joists are used for the direct support of roof and floor decks. They usually sustain only uniformly distributed loads on their top chords. A popular form for these is shown in Figure 7.56 with chords of cold-formed sheet steel and webs of steel rods. In addition, the chords also can be double angles, with the rods sandwiched between the angles at the joints.

Truss members can be designed to attach together in several different ways. The two most common methods are single plane, which usually calls for splay cutting or gusset plates (Figure 7.57A), and layer-on-layer in which the members are placed with their webs touching back to back (Figure 7.57B). Evaluate the truss connections for shear, moment, and axial forces due to the eccentricities of the connected members.

When selecting trusses, make sure they are made of certified steel. Certified steel provides the engineer or designer with a knowledge of the steel used in the truss and a base on which to make calculations. Thoroughly check and test the truss configuration to verify the calculations.

Calculating Truss Loads

Every truss is unique in design because of governing codes, spacing, span, slope, and load. Live loads are produced both during and after construction by movable objects other than wind, snow, seismic, and dead loads. Frequently used values for live loads are shown in Table 7.1.

FIGURE 7.56 Short-span, open-web steel joist.

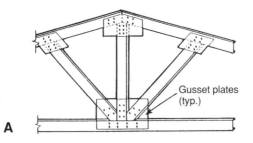

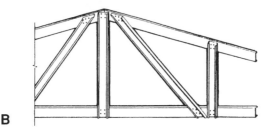

FIGURE 7.57 Two most common methods: (A) single plane, and (B) layer-on-layer.

TABLE 7.1 Roof Live Loads

Loaded area for any structural member (ft²)			
Roof slope	0–200	200–600	Over 600
25/12 to 4/12	20	16	12
4/12 to 12/12	16	14	12
Over 12/12	12	12	12

Two important factors must be considered when designing a truss.

Snow Load. The value used for snow load is specified by the local building code. Consider special conditions that encourage heavier than normal accumulations, such as multiple gables, abrupt changes in roof elevation, etc.

Wind Load. This is the amount of load that is imposed on the structure as a result of wind in any horizontal direction. The wind-load value is determined by the local building codes. The action of the wind on a structure causes forces to be applied on all exposed surfaces. Determining the value of the wind load for a roof or wall requires the use of coefficients. The sample calculations refer to the coefficients in Figure 7.58 and Table 7.2.

To obtain the value for these coefficients, refer to the local building codes. Typical values for roof-wind coefficients (RCOEF) and wall-wind coefficients (WCOEF) are 1.2 and 1.1 respectively. These apply only to enclosed buildings with a roof slope that is less than 10 degrees.

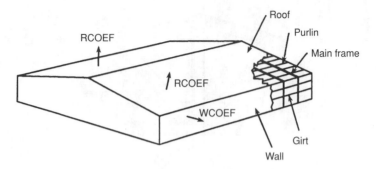

FIGURE 7.58 Forces acting on the eccentricity of connecting members.

TABLE 7.2
Coefficients of Roof Design

Surface	Member type	Symbol
Roof	Purlins	RCOEF
Walls	Girts	WCOEF

Give consideration to the type of truss, quantity, and location of the connection points, types of connections, bridging, trucking, handling, and all aesthetic concerns. Trusses, in fact, are rigid frames and therefore react differently under load than most framing methods. These differences must be taken into account within the frame, as well as between the frame and the structure. They need to be analyzed completely, uniformly, and accurately.

As we have already seen, roof trusses, in addition to their own weight, support the weight of roof sheathing, roof beams or purlins, wind loads, snow loads, suspended ceilings, and sometimes cranes and other mechanical equipment. Provisions for loads during construction and maintenance often need to be included. Distribute all applied loading to the truss in such a way that the loads act at the joints. In addition, include the required live-load roof designs specified by the local building codes.

When snow is a potential problem, base the load on anticipated snow accumulation. Otherwise the specified load is intended essentially to provide some capacity for sustaining loads experienced during the construction and maintenance of the roof. The basic required load usually can be modified when the roof slope is of some significant angle and based on the total roof-surface area supported by the structure.

When planning cuts, try to square-cut the studs or use them as they come from the roller former. Remember lightweight steel is not the most user-friendly material when it comes to compounds or miter cuts. A hip rafter can be fastened into place with just a square-and-cut stud. In all cases, however, end blocking or bridging for rafters and trusses usually is recommended over supports when the bearing ends are not otherwise restrained from rotation. Provide web stiffeners, as determined by the design, at the bearing points and at points of concentrated loads. The web stiffeners can be a small stud, track, or angle section designed to carry axial loads and cut to sit squarely against the supports.

In seismic and hurricane-prone areas (Figure 7.59), consider using the truss and rafter ties shown in Figure 7.60. These ties help hold a building together during earth movements and very high winds. It is important to securely tie together any steel-framed structure. For instance, the lower chords of fully loaded trusses sometimes settle. To compensate for this,

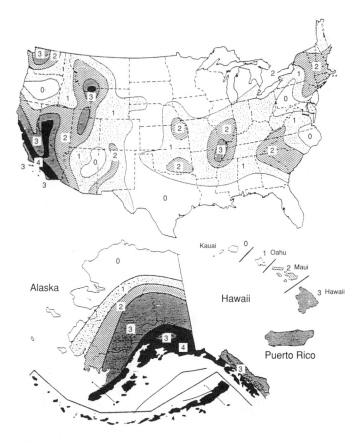

FIGURE 7.59 Zones of probable seismic intensity; Zone 4 is most probable.

introduce a slight camber into the lower chord (Figure 7.61). The amount of camber depends on various factors, including the span, load, and slope of the roof. The flatter the roof, the greater the stresses on it.

When there are load problems, use steel gusset plates to facilitate the connections. These plates have loads applied from several directions and again must be analyzed completely. Connections can be accomplished with screws or welds. The choice is between the designer and the fabricator.

Fastener application in roof systems should be located at the:

- Subfascia to the rafter
- Rafter to rafter
- Collar tie to the rafter
- Bridging to the rafter
- Rafter to the ceiling joist
- Gusset to the rafter

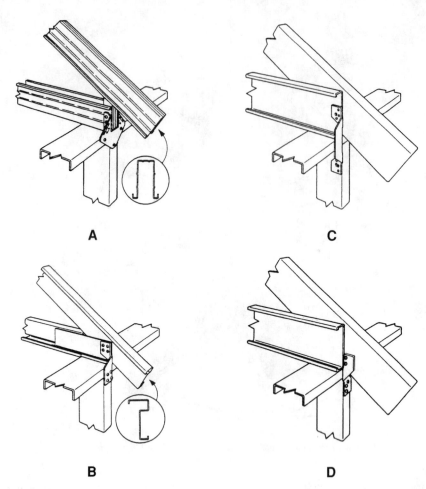

FIGURE 7.60   Truss and rafter ties.

- Kingpost to the rafter
- Truss web to the rafter
- Bracing to the rafter
- Rafter to the web stiffener

COMPARING TRUSS AND STICK-FRAMED ROOFS

Which is better: a stick-built or truss-constructed roof? The advantages to a stick-framed roof are:

- Roof framing materials are delivered with wall and floor-framing materials
- Installation does not require the use of a crane or special equipment

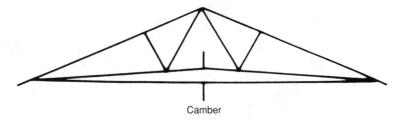

Camber

FIGURE 7.61   Camber is introduced into lower chord to allow for sag.

- Volume space in the attic is available for storage
- Future expansion into living space is possible and economical
- Future change to a full cathedral ceiling is possible

Some of the disadvantages to stick-framed roof construction are:

- The interior bearing is supported by the addition of a center footing or girder
- An interior bearing wall must be installed to support ceiling joists
- Field labor requires the knowledge needed to use a framing square

The advantages to truss-type roof designs are:

- No center bearing walls needed to support the trusses
- Labor in the field is reduced and simplified

NOTE: The interior walls typically are installed at the approximate center bearing to separate space, but no credit has been given for this wall in our comparison.

Some of the problem areas with truss-framed roofs are:

- Installation requires the coordination and cost of a crane
- Damaged material is more difficult to replace than in-stock stick items
- Design phases require additional engineering and coordination
- Offsite fabrication requires the coordination of an additional supplier
- Cost of future expansion into the roof system (i.e., cathedral ceilings) is prohibitive
- Volume space in the attic is not available for storage

ASSEMBLING TRUSS JIGS

Large-volume residential steel builders often build their own trusses in a shop and then truck them to the job site (Figure 7.62). Smaller builders usually buy trusses from local fabricators or distributors. Steel trusses must be built to rigid specifications that have been established by national building codes.

Assembly hardware is now available and automated setups are available to gang-fasten trusses. The designs for the latter are determined from data input into a computer. The computer software available for each type of truss makes it possible to obtain detailed con-

FIGURE 7.62 Trusses are built in shops and trucked to the site.

struction data. Information such as span, slope, type and gauge of metal, and type of fasteners is input into the computer for a specific type of truss. Output data includes lengths of members, positions of splices, positions of webs, and angles of cuts.

A jig can be assembled to rapidly position and fasten members after the design of the truss is known. Take cutting angles from a pattern truss and draw up a cut-list. While suitable for use in any type of construction, trusses are best in buildings with a rectangular design.

Trusses are ready to install as soon as they are unloaded at the site. Unload them in an upright position and carry them at their normal points of support to the storage area. Always store trusses upright and support them only at their bearing points.

Once all trusses are lifted into place and fastened, apply the sheathing or plywood.

APPLYING ROOF SHEATHING

Once the steel roof frame is complete, regardless of design, install the roof sheathing. For asphalt and other composition materials, solid sheathing is required (Figure 7.63). Wooden shingles and shakes installed in damp climates can be installed on spaced sheathing. Possible sheathing materials include shiplap and common boards, plywood, and nonwood materials. Regardless of the material used, all sheathing should be smooth, securely fastened, and provide a suitable base for roofing nails and fasteners.

When roofs are flat or of low pitch, install sheathing boards diagonally to resist racking. Although all roof-sheeting lumber should be seasoned, it is especially important that the boards are well seasoned when asphalt shingles are used. The use of poorly seasoned lumber might cause the asphalt to buckle. The maximum moisture content should not exceed 12 percent.

Plywood is ideal for sheathing. The large-size panels add rigidity to the roof, they are easily installed, and they provide a sound base for the roof covering. Install plywood with the face grain perpendicular to the rafters (Figure 7.64). In areas where damp conditions

FIGURE 7.63 Inside view of solid sheathing required for asphalt and other composition materials.

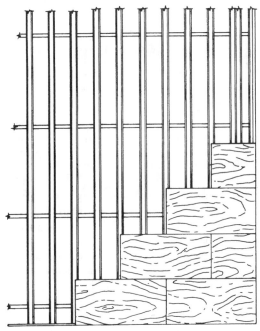

FIGURE 7.64 Install plywood with the face grain perpendicular to the rafters.

exist, use standard sheathing with an exterior glueline. For all roofs with exposed, open soffits, apply an exterior glueline to the perimeter panels. Be sure that all end joints occur over the center of a rafter.

Nearly all roof sheathing problems are due to incorrect installation. Following these . simple construction steps provides for the best performance and minimizes callbacks.

• Always check for a level fastening surface. This can be done with a piece of lumber (6 to 10 feet long) or a long carpenter's level. Shim trusses or rafters as necessary to provide a level fastening surface.

• Provide roof ventilation according to the building codes. The minimum net free ventilation area usually is 960 square inches for each 1000 square feet of ceiling area. When vents are located at eaves or soffits, near peaks, or along ridges for maximum air flow, the free-vent area can be reduced. The minimum standard is 480 square inches per 1000 square feet.

• Vent exhaust air from kitchens and baths to the outside with vent pipes that run through the roof cavity or attic to roof ventilators. Do not vent exhaust air directly into the roof cavity or attic.

• Additional air flow can be had by installing baffles between the framing and/or under the roof sheathing at the eaves to ensure that ceiling or roof insulation does not block ventilation paths. For vaulted or cathedral-ceiling construction, provide free ventilation paths from the eaves to the ridge between all rafters.

• Fasten panels with at least panhead screws spaced on maximum 6-inch centers at supported panel ends and edges. At the intermediate supports, fasten the panels 12 inches on center. High-wind areas might require more fasteners. Fasteners should be nominally ⅜ inch from the panel ends and ⅜ inch from the panel edges.

In addition to these guidelines, here are some additional recommendations to consider when installing roof sheathing over a steel-framed roof.

• Position the panel carefully. Use temporary fasteners at the corners, if needed, to square the panel on the framing.

• Install the fasteners at one panel end.

• Remove temporary fasteners at the corners.

• Install intermediate fasteners, starting at the panel edge. Use a chalkline or straight edge to align the fasteners on the framing. Fasten the panels in rows across the panel width. Continue this sequence along the length of the panel. This procedure keeps internal stress from accumulating in the panels.

• Stand on the panel near the fastener location to ensure contact with the frame when the fasteners are driven. Drive the fasteners flush with the panel surface.

• For improved performance, consider thicker sheathing panels, panel edge clips, or panels with tongue-and-groove edges.

• When installing the panels, a ⅛-inch space between adjacent panel ends and edge joints is recommended, unless the panel manufacturer indicates otherwise.

• Check the building code requirements for installing panel edge clips. Edge-clip requirements depend on the relationship of the panel span rating to the actual distance between the roof framing.

• Cover the sheathing with underlayment felt as soon as possible to minimize exposure to weather, unless otherwise recommended by the sheathing manufacturer.

When installing sheathing on the roof, it is important to leave a space between the sheets to permit expansion. Waferboard and oriented-strand board (OSB) are made under extreme heat and pressure. They are palletized in this process and thus never have the opportunity to absorb moisture. Once installed on the building, allow the panels to absorb moisture and expand fully. Most shingle manufacturers do not recommend installing shingles until a week after the sheathing is installed. If the shingles are installed too soon, or if there is insufficient space between the sheets, visible swollen edges under the shingles, or *ridging*, can occur. H-shaped ply-clips are used on the edge of the sheathing between purlins and are recommended primarily for edge support.

FIGURE 7.65   Steel-framed roofs can be finished with asphalt shingles, tile, wooden shakes, or metal.

FINISHING THE ROOF

The steel-framed roof can be finished with asphalt shingles, tile, wooden shakes or shingles, or metal (Figure 7.65). Metal offers many advantages to a steel-framed home.

As we have seen, metal is a predictable and stable product by nature. The forces of nature on metal roof panels are well understood from years of accumulated research. Forces, such as those brought on by wind loading and thermal activity, are considered when these systems are evaluated. Unlike other single-ply technologies, such as rubber, a metal panel is not affected by ultraviolet rays, which are one of the leading sources of roof degradation.

Advances in metal coating and finishing technologies have improved significantly the life-cycle cost benefit of metal roofs. Metal roofing also is extremely puncture-resistant, which is a real benefit to today's building owner who typically is looking for long-term, low-maintenance solutions. Actually, durability is one reason for metal's growing popularity. Many standing-seam roof systems have earned a Class-90 wind-uplift rating from Underwriters Laboratories, Inc. (UL), which is the highest rating in the industry. Some systems also can carry a Factory Mutual Class A fire rating due to their noncombustible surface. Both ratings can help the building owner substantially reduce building insurance rates.

Metal roofing is extremely versatile and colorful, can be incorporated into any design concept, and is compatible with all types of building materials. Metal roofing accentuates masonry, wood, stucco, glass-curtain walls, marble, and granite. The various profiles available in metal panels make the adaptation from large projects to small ones very easy. Eighteen- to 24-inch-wide panel configurations are best suited for large commercial structures, while 8- to 12-inch-wide profiles are most often seen on smaller structures such as residential or small commercial buildings.

CHAPTER 8
INTERIOR FRAMING DETAILS

Steel framing frequently is installed to meet special construction applications. For example, the framing might be designed to improve decorative effects, add moisture resistance, improve thermal insulation, reduce sound transmission, provide special construction features, or meet fire-code regulations. Many of the ideas and suggestions put forth in this chapter can be used in remodeling applications.

HANGING DRYWALL

Drywall and steel-framed construction seem to go hand in hand. Both are very versatile. The first seven chapters of this book discussed the advantages to and use of steel framing. The drywall installer who applies the finishing material over the steel framing can work with a variety of sizes and types of gypsum board. The standard gypsum board is used primarily as an interior wall and ceiling surface for finishes such as paint, wallpaper, and other wall coverings. It also is available prefinished in colorful textures, vinyl, or fabric-surfaced panels.

In addition, there are gypsum boards on the market that can be used for ceramic, metal, and plastic tile; as area separation walls between occupancies; for exterior soffits; as an underlayer on exterior walls (see Chapter 6); and to provide fire protection to structural elements. Many of these gypsum boards can be purchased with aluminum-foil backing, which provides an effective vapor retarder for exterior walls when it is applied with the foil surface against the framing.

Table 8.1 gives the General Service Administration's (GSA) federal specification designation of gypsum materials that can be used in conjunction with steel framing.

TABLE 8.1 Federal Specification Designations

Type	Grade	Class	Form	Style
TYPE I Lath	R-Regular core X-Fire-retardant core	1-Plain face	a-Plain back b-Perforated c-Foil back	1-Square edge 5-Round edge
TYPE II Sheathing	R-Regular core W-Water-resistant core X-Fire-retardant core	2-Water-resistant surface	a-Plain back	1-Square edge 2-V-tongue and groove edge
TYPE III Wallboard	R-Regular X-Fire-retardant core	1-Plain face 3-Predecorated surface	a-Plain back c-Foil back	1-Square edge 3-Taper or recess edge 4-Featured joint edge 6-Taper, featured edge
TYPE IV Backer board	R-Regular core X-Fire-retardant core	1-Plain face	a-Plain back c-Foil back	1-Square edge 2-V-tongue and groove edge 3-Taper or recess edge
TYPE V Formboard	R-Regular core	4-Fungus-resistant surface	a-Plain back	1-Square edge
TYPE VI Veneer plaster base	R-Regular core X-Fire-retardant core	1-Plain face	a-Plain back c-Foil back	1-Square edge 3-Taper or recess edge 6-Taper, featured edge
TYPE VII Water-resistant backing board	R-Regular core W-Water-resistant treated core X-Fire-retardant core	2-Water-resistant surface	a-Plain back	1-Square edge 3-Taper or recess edge

*As an aid in the identification of gypsum wallboard products, above are the classifications as set forth in the Federal Specifications SS-L30D.

APPLYING WALL PANELS

Once the steel-framing studs for a wall or partition are in place, the drywall panels can be installed. The gypsum board, for applications over steel studs, can be applied horizontally or vertically in single or double layers. With double-layer construction, the face or finishing layer can be prefinished gypsum wallboard adhesively applied to the base layer or ply.

While drywall application normally is *not* in a steel framer's job description, knowing how it is done gives the framer a better overall understanding of the drywall mechanic's tasks.

Using Single-Layer Applications

Apply the gypsum-drywall panel with the length parallel or at right angles to the studs (Figure 8.1). Cut the wallboard to size following the steps illustrated in Figure 8.2. Center the abutting ends or edges over the stud flanges. Locate all the attaching screws on 12-inch centers when the framing is on 24-inch centers and on 16-inch centers when the metal framing is on 16-inch centers or less.

Attach the gypsum drywall to the studs with type-S drywall screws using an electric drywall screwdriver with a No. 2 Phillips bit (Figure 8.3). For vertical panel application when the studs are on 24-inch centers, erect the drywall on one side of the partition and screw it to the studs at the vertical wallboard joints. Drive the first row of screws at a panel joint into the face, or *flange*, of the stud adjacent to the stud web. Apply the drywall panels to the entire side of the partition in this manner.

For the opposite side, cut the first wallboard panel so that the joints are staggered. Fasten this and all succeeding drywall panels to all studs on this side. When the partition face is complete, return to the first side and finish screwing the drywall to all the intermediate studs.

For a single-layer adhesive application, prebow the panels (Figure 8.4) and attach them vertically to the studs using ⅜-inch continuous adhesive beads applied to the face of the studs. Apply one bead to the intermediate studs and two beads to the studs that occur at the panel joints. Secure the panel at the top and bottom with 1-inch type-S screws spaced on 16-inch centers. Impact the panel along each stud to ensure good contact at all points.

Using Double-Layer Applications

Two-ply or double-layer drywall construction with steel studs offers some of the best performances in both fire and sound resistance. This construction method can achieve up to a

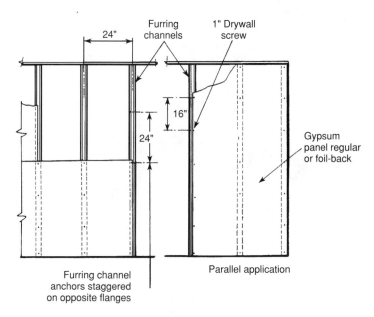

FIGURE 8.1 Perpendicular (vertical) and parallel (horizontal) application of gypsum drywall panels.

A

B

C

FIGURE 8.2 (A) Cut the gypsum board by scoring it with a utility knife against a T-square; (B) Snap it back; (C) Cut the back paper with the same knife and separate the sections; (D) Snap the gypsum board for a clean break.

D

FIGURE 8.2 *(Continued)*

FIGURE 8.3 Installing drywall panel with an automatic screwdriver.

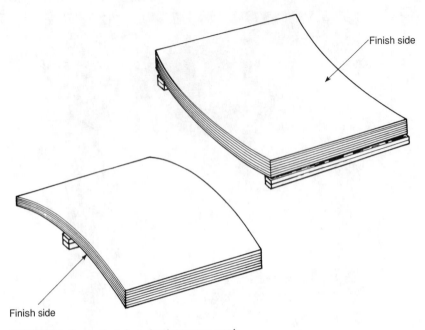

Finish side

Finish side

FIGURE 8.4   Two methods for prebowing gypsum panels.

2-hour fire rating and a 55-sound transmission class (STC) sound rating (Figure 8.5). These economical, lightweight partitions are adaptable as party walls or corridor walls in virtually every type of new construction.

To attach each of the double-layers with screws, fasten the face ply to the metal studs with type-S screws spaced on 12-inch centers. Use 1⅝-inch screws for ½- and ⅝-inch-thick gypsum board. As a rule of thumb, screws should be at least ⅜ inch longer than the total thickness of the material to be attached to the steel studs.

For double-layer adhesive-laminated construction, attach the base layer with 1-inch screws spaced on 8-inch centers at the joint edges and on 12-inch centers in the field. Apply the face layer vertically and spread the specified setting-type joint compound or ready-mixed joint compound on the back side. Stagger the joints approximately 12 inches and fasten them to the base layer with 1½-inch laminating screws. Drive the screws approximately 2 feet from the ends and on 4-foot centers in the panel field, 1 foot from the ends and on 3-foot centers along a line 2 inches from the vertical edges. Temporary shoring or support, placed 16 to 24 inches on center until the adhesive dries, can be used in place of screws.

For double-layer, laminated, nonrated construction, attach the base layer with 1-inch type-S screws spaced 16 inches on center at the joint edges and in the panel field. Apply the laminating adhesive in strips to the center and along both edges of the gypsum face panel. Apply the strips with a notched metal spreader that has at least four ¼-×-¼-inch notches spaced on maximum 2-inch centers. Position the wall panels vertically, press them into place with firm pressure to ensure a bond, and fasten the top and bottom as required. For ceiling panels, space fasteners on 16-inch centers along the edges and ends. Install one permanent field fastener per framing member at the midwidth of the panel.

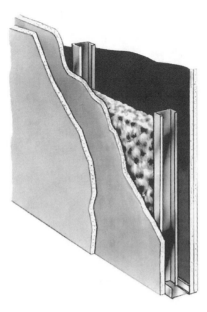

FIGURE 8.5 Double-layer installation onto metal studs with insulation installed in between.

For resilient construction, apply the wallboard with its long dimension perpendicular to the resilient channels and fasten with 1-inch type-S screws spaced on 12-inch centers along the channels. When the channel resiliency makes the screw placement difficult, use the next longer screw, but do not drive the screw directly over the stud (see Chapter 9).

Wallboard can be applied directly over masonry using metal furring strips (Figure 8.6). Adjustable brackets make it easy to install the furring strips over irregular masonry walls (Figure 8.7).

INSTALLING CEILING FRAMING

Three types of screw-attached, steel furring assemblies can be used to attach gypsum-drywall board to ceilings:

- Suspended furring ceiling channel
- Resilient suspended furring channel
- Direct-screw furring studs

All three assemblies are suitable for fastening to the lower chord of steel joists or for carrying channels in suspended ceiling constructions. Secure direct-screw furring channels and resilient furring channels with tie-wires or furring clips (see Chapter 2). Attach metal screw studs with wires. Table 8.2 indicates the maximum spans recommended for each type of furring with ½- or ⅝-inch gypsum panels.

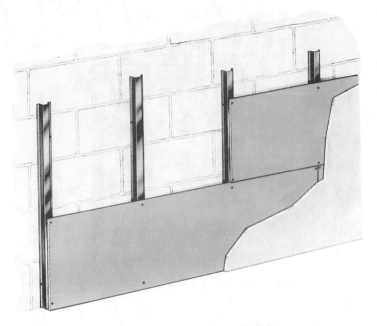

FIGURE 8.6   Gypsum board applied to masonry surfaces with metal furring.

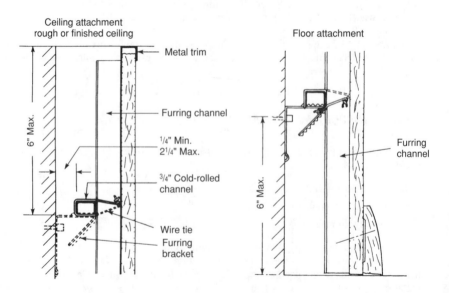

FIGURE 8.7   Adjustable brackets make installation of furring easy over irregular masonry walls.

TABLE 8.2 Furring Member Spacing

Furring member	Furring member spacing	Furring member spacing	Furring member spacing
0.0179" min. base steel	24" O.C.	16" O.C.	12" O.C.
Resilient furring channel	2'0"	2'0"	2'0"
Furring channel	4'0"	4'6"	5'0"
1⅝"	5'0"	5'6"	6'0"
2½"	6'0"	6'6"	7'0"
3⅝"	8'0"	8'6"	9'0"
4"	8'6"	9'0"	9'6"
6"	8'6"	9'0"	9'6"

Framing Suspended Furring Channels

The furring channel ceiling, framed to receive drywall, is one of the most popular types of ceiling used today in commercial and residential structures (Figure 8.8). To erect the suspended ceiling grille, start by spacing 8-gauge hanger wires on 48-inch centers along the

FIGURE 8.8 Commercial application of a suspended-ceiling system.

carrying channels and within 6 inches of the ends. In concrete, attach the anchor hangers to reinforcing steel by loops that have been embedded at least 2 inches or by approved inserts. For steel-framed construction, wrap the hanger around or through the beams or joists.

Install 1½-inch carrying channels on 48-inch centers and within 6 inches of the walls. Position the channels for the proper ceiling height, then level and secure by saddle-tying the hanger wire along the channels as shown in Figure 8.9. Provide a 1-inch clearance between the runners and the abutting walls and partitions. At the channel furring splices and the interlock flanges, overlap the ends 6 inches and secure each end with double-strand, 18-gauge tie-wire.

Erect the steel furring channels at right angles to the 1½-inch carrying channels (Figure 8.10). Space the furring to within 6 inches of the sidewalls. Provide a 1-inch clearance between the furring ends and the abutting walls and partitions. Attach the furring channels to the 1½-inch channels with furring channel clips installed on alternate sides of the carrying channel. Saddle-tie the furring to the channels with double-strand, 18-gauge tie-wire when the clips cannot be alternated. At the splices, nest the furring channels with a 6-inch overlap and securely tie each end with double-strand 18-gauge tie-wire (Figure 8.11).

Where required for direct-suspended fire-rated assemblies, install the double furring channels to support the gypsum-board base ends and back-block with a gypsum board strip. When staggered end joints are not required, use control joints.

Apply the gypsum panel with its long dimension at right angles to the furring channels. Locate the panel's butt joint over the center of the furring channels. Attach the drywall boards with 1-inch self-drilling drywall screws on 12-inch centers in the panel field and on 8- to 12-inch centers at the butt joints, located not less than ⅜ inch or more than ½ inch from the edges.

Framing the Exposed-Grid Suspended System

Predecorated gypsum ceiling panels can be laid in a conventional exposed-grid system (Figure 8.12). Various grid patterns are shown in Figure 8.13. Because these lay-in panels

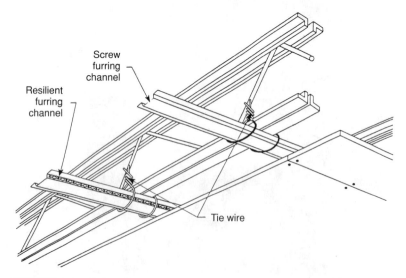

FIGURE 8.9 Installation of hanger wires when using furring channels in a suspended ceiling system.

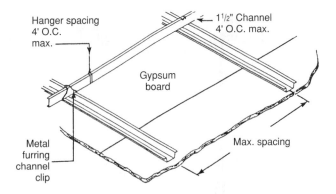

FIGURE 8.10 Installing the gypsum panels to metal furring channels.

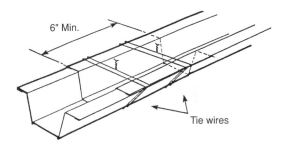

FIGURE 8.11 A furring-channel splice joint.

FIGURE 8.12 Typical example of an exposed-grid system.

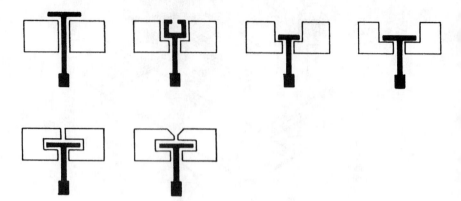

FIGURE 8.13   Various grid lay-in patterns are available. Rigid cores prevent sagging or bowing.

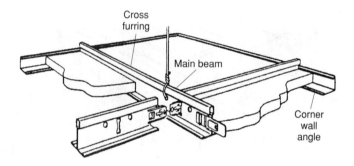

FIGURE 8.14   Installing an exposed-grid system.

have a rigid gypsum core, they resist sagging and warping and do not require clips to off-set bowing (Figure 8.14). Compliance with the following cautions and restrictions ensures maximum performance from these gypsum ceiling panels.

- Support each panel on all four edges.
- Do not install the panels in areas exposed to extreme or continuous moisture.
- For exterior applications, protect the grid panels from direct exposure to weather, water, and continuous high humidity. Under no circumstances should water from condensation or any other source come in contact with the back of the panels.
- Install a control joint when the surface area requires it (Figure 8.15).
- Overlaid insulation might cause excessive panel deflection and is not recommended where high humidity is likely to occur.
- Provide cross ventilation in unheated areas or in the enclosed space above the ceiling panels.
- Use the standard $^{15}\!/_{16}$-inch exposed-tee grid or an environmental-type grid in severe conditions for the suspended-grid system.

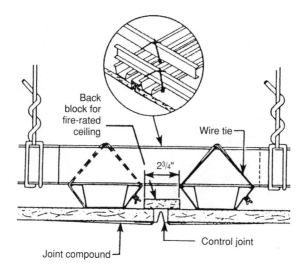

Back block for fire-rated ceiling

Wire tie

2³/₄"

Joint compound

Control joint

FIGURE 8.15 Installing a control joint in an exposed-grid system.

- Extreme lighting conditions might distort the appearance of the texture.
- Remove common dirt and stains on the vinyl surface with mild soap or detergent in lukewarm water. Use a light scrubbing action with a cloth, sponge, or soft brush.

Installing Direct-Screw Furring

When installing a direct-screw system, attach the gypsum board to the wall angles at the perimeter of the area. Once the ceiling height is determined, mark this height with a chalk-line or the beam of a laser alignment tool (Figure 8.16). Then install the wall angles along these lines. Locate crossfurring channels within ⅛ inch of the walls without wall angles and within 8 inches of the panel end joints.

Erect the metal furring at right angles to the bar joists or other structural members. As an alternative, use the studs as furring. Saddle-tie the furring channels to the bar joists with double-strand 18-gauge tie-wire at each intersection. Provide a 1-inch clearance between the furring ends and the abutting walls and partitions. At the splices, nest the furring channels with at least an 8-inch overlap and securely wire-tie each end with double-strand 18-gauge tie-wire. The maximum allowable spacing for the metal furring channel is 16 to 24 inches, as required for ½- or ⅝-inch-thick gypsum board.

For bar joist spacing up to 60 inches, use the steel studs as furring channels. Wire-tie the studs to the supporting framing. Position the 1⅝-inch studs with the open side up. Position the larger studs with the opening to the side. See Table 8.3 for stud spacings and limiting spans.

Install gypsum board directly to the furring using basically the same method used to install suspended ceilings. Apply the drywall panel with its long dimensions at right angles to the channels. Locate the board's butt joints over the center of the furring channels. Attach the panels with 1-inch self-drilling or self-tapping screws on 12-inch centers. The screws should not be less than ⅜ inch nor more than ½ inch from the edges.

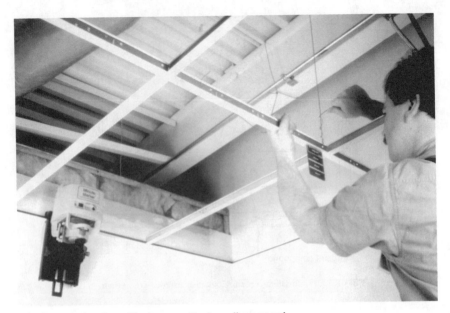

FIGURE 8.16 Leveling ceiling hangers with a laser-alignment tool.

TABLE 8.3 Limiting Span for Steel Furring Members

Type furring member	Member spacing (in.) (O.C.)	Single-layer panels (2.5-psf max.)		Double-layer panels (5.0-psf max.)	
		1-span	3-span	1-span	3-span
25" stud (hemmed)	16	5'9"	7'1"	4'7"	5'8"
	24	5'0"	6'2"	4'0"	4'11"
20" stud (hemmed)	16	6'11"	8'6"	5'5"	6'9"
	20	6'0"	7'5"	4'9"	5'11"
25" stud	16	7'2"	8'10"	5'8"	7'0"
	24	6'3"	7'9"	5'0"	6'2"

To prevent an objectionable sag in new panel ceilings, follow these guidelines to determine the appropriate weight for the overlaid, unsupported insulation.

- Do not exceed 1.3 pounds per square foot (psf) for ½-inch-thick panels with a frame spacing on 24-inch centers
- Do not exceed 2.2 psf for:
  ~½-inch-thick panels with a frame spacing on 16-inch centers
  ~½-inch interior gypsum ceiling board with a frame on 24-inch centers
  ~⅝-inch panels with a frame on 24-inch centers
- Do not overlay ⅜-inch-thick panels with unsupported insulation.

Install a vapor retarder in exterior ceilings, and properly vent the plenum or attic space.

FRAMING WALLS WITH CURVED SURFACES

Drywall panels can be formed to almost any cylindrical or curved surface (Figure 8.17). Panels are applied either dry or wet, depending on the radius of the desired curvature. To prevent flat areas between the framing, shorter-bend radii require closer-than-normal stud and furring spacing. The panels usually are applied horizontally, bent gently around the framing, and then securely fastened to achieve the desired radius. Table 8.4 supplies the minimum curvature radii for situations when boards are applied dry. These dry-bending radii are suitable for many curved-surface applications.

TABLE 8.4 Minimum Radius of Curvature When Boards are Applied Dry

Board thickness	Board applied with long dimension perpendicular to framing	Board applied with long dimension parallel to framing
Inches	Feet	Feet
½	20[1]	—
⅜	7½	25
¼	5[1]	15

[1]Bending two ¼-inch pieces successfully permits radii shown for ¼-inch gypsum board.

FIGURE 8.17 Drywall panels can be formed to almost any cylindrical or curved surface.

TABLE 8.5 Minimum Bending Radii of Wet Gypsum Board

Board thickness in inches(1)	Minimum radius in feet	Length of arc in feet(2)	Number of studs on arc and tangents(3)	Approximate stud spacing in inches(4)	Maximum stud spacing in inches(4)	Water required per panel side in ounces(5)
¼	2	3.14	8	5⅜	6	30
¼	2.5	3.93	9	5⅞	6	30
⅜	3	4.71	9	7$\frac{1}{16}$	8	35
⅜	3.5	5.5	10	7$\frac{5}{16}$	8	35
½	4	6.28	8	10¾	12	45
½	4.5	7.07	9	10⅝	12	45

(1)For gypsum board applied horizontally to a 4-inch-thick partition

(2)Arc length = 3.14R/4 (for a 90 degree arc)

(3)No. studs = arc length/max. stud spacing + 1 (rounded up to the next whole number)

(4)Stud spacing = arc length/no. of studs (measured along outside of runner)

(5)Wet only the side of the panel that will be in compression.

Shorter radii, such as those given in Table 8.5, can be obtained by thoroughly moistening the face and back papers of the board with water, stacking the boards on a flat surface, and then allowing the water to soak into the core for at least one hour. When the board dries, it regains its original hardness.

Figure 8.18 shows a curved-wall surface formed with two layers of gypsum drywall on a metal frame. To build the frame, cut one leg and the web on the top and bottom steel runner at 2-inch intervals for the length of the arc. Allow for a 12-inch uncut steel runner at each end of the arc. Bend the runners to the uniform curve of the desired radius, up to a maximum 90-degree arc. To support the runner cut leg, clinch a 1-inch-wide 25-gauge steel strip to the inside of the leg using metal lock fasteners. Select the runner size to match the steel studs. As shown in Figure 8.19, attach the steel runner to the structural elements at the floor and ceiling with suitable fasteners.

Position the studs vertically, with their open sides facing in the same direction, and engage the studs in the floor and ceiling runners. Begin and end each arc with a stud. Space intermediate studs equally as measured on the outside of the arc. Secure the steel studs to

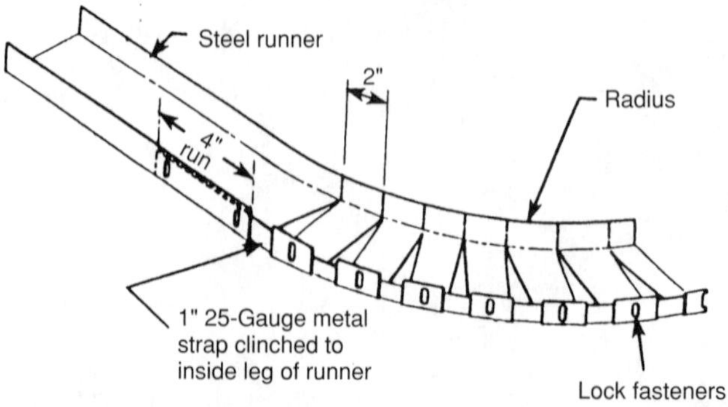

FIGURE 8.18 An example of the way bottom and top runners are formed for curved-board applications.

FIGURE 8.19 Fastening the bottom runner.

the runners with ⅜-inch type-S panhead screws. Secure the wooden studs with suitable fasteners. On tangents, place the studs on 6-inch centers and leave the last stud freestanding. The shorter the radius of the bend or curve, the closer the spacing of the framing members.

Apply the panels horizontally with the wrapped edge perpendicular to the studs (Figure 8.20). On the convex side of the partition, attach one end of the gypsum board to the framing with nails or screws. Progressively push the gypsum board into contact with each subsequent framing member and fasten. Work from the fixed end to the free end. Tightly hold the gypsum board against each supporting framing member when driving the fasteners (Figure 8.21).

On the concave side of the partition, apply a stop at one end of the curve to restrain one end or edge of the gypsum board during installation. Apply pressure to the unrestrained end or edge of the gypsum board to force the field of the gypsum board into firm contact with the framing. Fasten the gypsum board by working from the stopped end or edge. Tightly hold the gypsum board against the supporting framing when driving the fasteners.

FRAMING ARCHWAYS

Wallboard can be applied to the inner face of almost any archway (Figure 8.22). Precut the gypsum board to the proper length and width. For short radii, either moisten the gypsum board or score across the width of the back paper with parallel score marks spaced at ½-, ¾-, or 1-inch intervals, depending on the radius. After making these *backcuts*, make the core at each cut so that the board can be bent to the curvature of the arch. Tightly hold the board against the supporting framing when driving the fasteners.

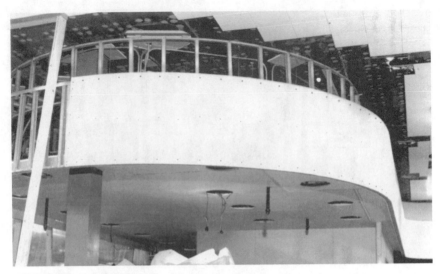

FIGURE 8.20 Typical interior soffit.

FIGURE 8.21 Installing a gypsum board to studs forming a curved-surface layout.

FIGURE 8.22 Typical drywall doorway arch.

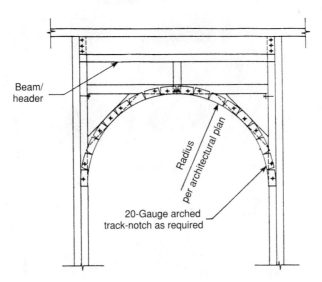

FIGURE 8.23 Typical method for constructing a doorway arch using steel framing.

The typical construction method for a doorway arch uses steel framing as shown in Figure 8.23. Plastic forms or inserts also are available that offer a variety of circular or curved shapes for use with drywall surfaces (Figure 8.24). These are easy to install and result in a perfect surface or arch (Figure 8.25).

A ¼-inch-thick high-flex gypsum board produced by some manufacturers is made specifically for curves (Figure 8.26). It is used for both inside and outside radii, such as archways, columns, and curved stairways. Except for tight radii work, it is not necessary to wet the board. Check the manufacturer's specifications for the bending-radii limitations of this specialty gypsum board.

INSTALLING PLASTER SYSTEMS

Other drywall finishing materials can be applied over a steel framing. Plywood, hardboard, wood, and vinyl predecorated panels are all examples of materials that can be installed over steel framing in basically the same manner as gypsum board. Of course, you should follow the manufacturer's directions to the letter when using these products.

Even plaster can be applied over steel framing. In these systems, a veneer plaster application that is ¹⁄₁₆- to ³⁄₃₂-inch thick and made of specially formulated gypsum plaster is applied over a gypsum base.

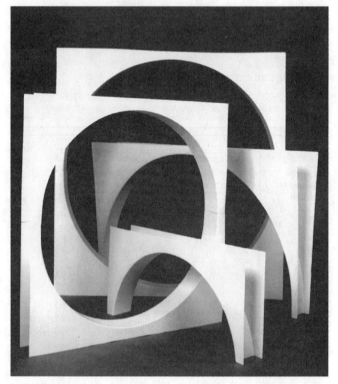

FIGURE 8.24 Various shapes or forms to make instant arches and circular openings in drywall surfaces.

FIGURE 8.25 Slip the shape into place and nail. Add edge trim.

Veneer plaster systems were developed during the 1960s and early 1970s. These systems took advantage of large-size gypsum panels to improve the speed of installation and provide more monolithic-appearing, harder, abuse-resistant surfaces. In addition, the overall plaster thickness was reduced from the standard ½ inch associated with conventional plaster to a mere ¹⁄₁₆ to ⅛ inch. This was achieved by using high-strength gypsum in the product formulations. The combination of less-thick and high-strength materials reduced drying time and provided a more serviceable finished surface.

Installing Partitions

In a single-layer application, set the steel studs in runners with one layer of ½-inch type-X (fire-rated) gypsum board attached with screws to 2½-inch studs on 16-inch centers. This assembly has a 1-hour fire rating and is suitable for interior partitions and corridor walls. This application, plus a sound absorbing insulation, offers a 50- to 55-STC rating.

A 2-hour fire rating with sound control suitable for party walls is available with a double-layer ½-inch type-X gypsum panel that is attached with type-S screws to 2½-inch minimum steel studs spaced on 24-inch centers.

Multilayer ½-inch type-X gypsum board and veneer-plaster assemblies offer 3- and 4-hour fire ratings and 62-STC sound control. When additional partition width is required,

FIGURE 8.26   Applying high-flex panels to sharp bends.

erect double rows of steel studs to provide chase walls with up to 20 ¾-inch net pipe-chase width. Up to 4-hour column fire protection also is available.

Using Conventional Plaster Systems

Conventional plaster systems also can be used with steel framing. The base is a solid gypsum-board lath that usually requires about 45 percent less basecoat plaster than metal lath. With convenient 16- and 24-inch widths and efficient 4- and 8-foot lengths, it is designed to quickly attach to steel framing with furring channels or suspended metal grille. Plaster metal lath also can be secured to steel framing as shown in Figure 8.27.

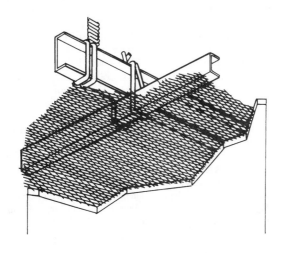

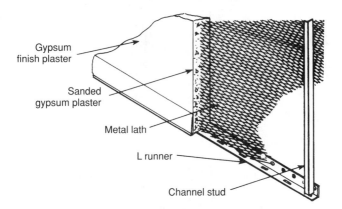

FIGURE 8.27 Using conventional plaster installation in place of drywall in steel-framing systems.

CONSTRUCTING STEEL-FRAMED PARTITIONS

There are several ways to construct interior steel-framed partitions. For example, the face panels of a solid wallboard partition can be ½ inch, ⅝ inch, or multiple laminations of regular or type-X gypsum board. The core also can be either 1-inch-thick gypsum core board or shaft liner in single or multiple layers. The laminating adhesive can be a joint compound for tape embedment or other laminating adhesive recommended by the gypsum-board manufacturer.

Each construction method starts by securing the metal runner or track to the floor with approved fasteners on 24-inch centers. Secure the ceiling runner in a manner appropriate for the ceiling construction. Use the L runner or track as both floor and ceiling runners in fire-rated construction.

Cut the gypsum panel ⅜ inch shorter than the wall height and install it vertically. Fasten the floor and ceiling runners with two type-S screws, each located 2 inches from each edge of the board panels (Figure 8.28). The perimeter relief at the ceiling and wall are shown in Figure 8.29. Invert the resilient channel to make it easier to attach to the base.

A solid laminated partition (Figure 8.30) generally is made from a 1-inch gypsum core board or shaft liner that is cut ⅜ inch shorter than the wall height. Install vertically and fasten it to the floor and ceiling runners with two 1⅝-inch type-S screws, each located 2 inches from each edge of the core-board panels. Apply a joint compound or adhesive evenly to the back side of the 4-foot-wide, ceiling-height wallboard face panels and vertically to the sides of the gypsum core board. Use a notched spreader that has ¼-×-¼-inch notches on 1½-inch centers. Stagger the wallboard joints 12 inches from the core-board joints and 24 inches from the joints on each side of the partition wall.

Screw the face panels on 12-inch centers to the bottom runners with 1⅝-inch type-S screws and secure the panels to the core board with 1⅝-inch type-S drywall screws spaced on 24-inch centers both horizontally and vertically to include screws along the vertical edges of each panel. Reinforce the drywall panel joints with paper tape and finish with a joint compound. Door frames should be vertical, plumb, and true. Securely anchor door frames to the floor, wall, and ceiling. The completed partition should be straight and plumb.

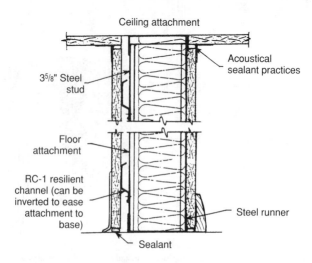

FIGURE 8.28 Typical steel-framing partition.

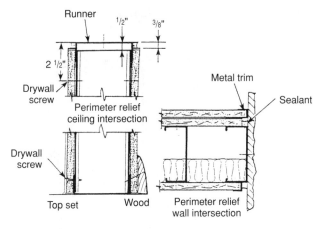

FIGURE 8.29 Partition perimeter relief details.

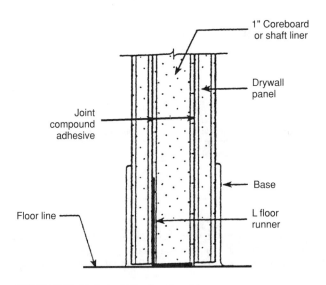

FIGURE 8.30 Typical solid-laminated partition.

The solid laminated partition can be adapted to any type of building for nonload-bearing partitions. Table 8.6 gives the allowable partition heights based on gypsum board applied to light-gauge steel that is acting as a composite section. The limiting heights that exceed those shown can be obtained by using deeper studs, spacing the studs closer together, using heavier-gauge studs, increasing the thickness of the gypsum board, or by adding an additional layer of wallboard. Of course, a semisolid partition can be constructed by omitting the solid gypsum-core panel and substituting insulation in its place (Figure 8.31).

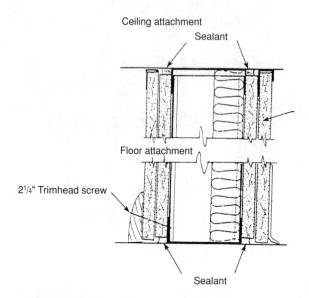

Ceiling attachment

Sealant

Floor attachment

2¹/₄" Trimhead screw

Sealant

FIGURE 8.31 Semisolid partition without gypsum core.

TABLE 8.6 Allowable Partition Heights

Stud spacing (inches)	Facing on each side	Stud depth (inches)			
		1⅝	2½	3⅝	4
		Height in feet and inches			
16	½"-one-ply	11'0"	14'8"	19'5"	20'8"
24	½"-one-ply	10'0"	13'5"	17'3"	18'5"
24	½"-two-ply	12'4"	15'10"	19'5"	20'8"

Figure 8.32A illustrates how to terminate a partition and Figure 8.32B shows how to install a wall control joint for partition runs that exceed 30 feet. A partition corner is detailed in Figure 8.33. The system used to attach a partition to the ceiling depends on the ceiling material and the method of installation (Figure 8.34).

Overcoming Tall-Wall Restrictions

Partitions that exceed 30 feet in height are considered tall. When these taller-than-normal partition heights are required, consideration must be given to several factors, including length restrictions on the manufacturing and shipping of steel studs, scaffolding design, and stud placement.

Place double studs back to back on 24-inch centers. The studs should be the maximum practical length so that the splice of one stud in each pair occurs within the outer ⅓ of the span. The splice of the other stud should occur at the opposite end. Attach the

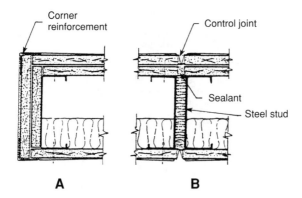

FIGURE 8.32 (A) Partition terminal, and (B) wall control joint.

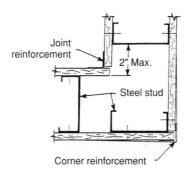

FIGURE 8.33 Partition corner joint.

studs back to back with screws approximately 4 inches on center. Attach each stud flange to the top and bottom runner with ½-inch type-S-12 screws so that each pair of studs has four screw attachments at each end. Attach 1½-inch 20-gauge V bracing to the stud flanges on each side on approximately 12-foot centers for stud alignment and lateral bracing.

For a 5-psf wind load, 20-gauge runner track is recommended. The fasteners should have a single shear and bearing capacity of 300 pounds. For a 10-psf wind load, attach 18-gauge runner track with fasteners that have a single shear and bearing capacity of 400 pounds.

Attaching Baseboards

Several methods can be used to attach baseboards. Contractors who choose to use nails either place a wooden plate beneath the steel runner or simply insert sections of wooden 2 × 4s between the studs in the floor-mounted runner. Others use trimhead screws or adhesive to attach the baseboard.

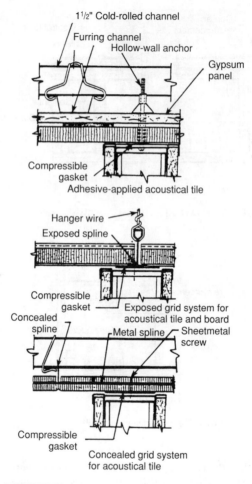

FIGURE 8.34 Attaching partition to ceiling.

Supporting Pipes and Protecting Wiring

Pipes can be fastened to steel studs and bracing with screw-attached straps. Wood can be inserted as a cushion to restrict movement and eliminate rattling. Wherever copper pipes or electrical wire passes through keyhole punch-outs in the stud web, use grommets or insulating sleeves to protect materials from sharp metal edges and galvanic action (Figure 8.35). Prevent galvanic action between copper tubing or pipe and metal framing by wrapping the copper pipe or tubing with insulation.

The attachment for a typical electrical connection is detailed in Figure 8.36. The installation of outlet boxes in steel-framed walls is shown in Figure 8.37.

FIGURE 8.35   Installing either rubber or plastic grommet.

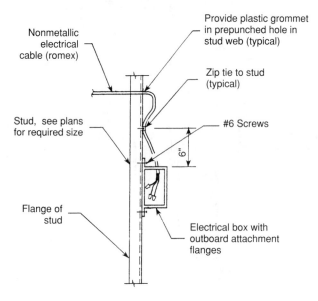

FIGURE 8.36   Typical attachment method for electrical boxes.

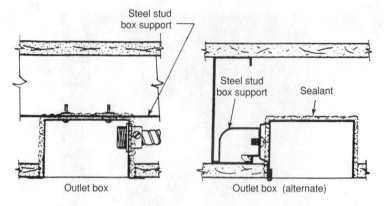

FIGURE 8.37 Methods of installing outlet boxes.

Using Chase or Hollow Walls

To simplify the installation of piping and ductwork and to reduce room-to-room noise, use a chase or hollow-wall system. The many benefits and possible savings offered by hollow-wall systems are inherent in the nature of the partition system. The clear, unobstructed space between partitions provides for pipes, ducts, and conduits (Figure 8.38). No fittings

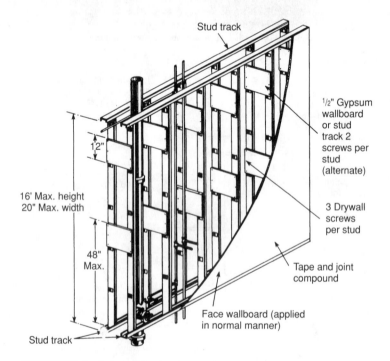

FIGURE 8.38 Typical metal-framing chase wall.

or costly adjustments are needed to accommodate vertical or horizontal piping runs or ductwork. Because the partition halves are separate units, they can be spaced far enough apart to conceal the installed fittings.

The three integrated metal parts of the partition are quickly and easily installed, which minimizes labor costs for this type of construction. The completed partition generally costs considerably less than masonry walls of the same thickness. No matter what the spacing is between the partition halves, the cost remains the same. Because it is ideal for concealing piping and for increasing sound resistance, the chase or hollow-wall system is useful in all types of structures where ductwork or plumbing needs to be hidden. Hollow-wall systems also accommodate all types of electrical fixtures, casings, electrical switches, and outlet boxes.

To assemble a chase or hollow-wall system, align two parallel rows of floor and ceiling runners according to the partition layout. Spacing between the outside flanges of each pair of runners must not exceed 24 inches. Follow the previous instructions for attaching runners. Then position the steel studs vertically in the runners with the flanges in the same direction and with the studs on opposite sides of the chase directly across from each other (Figure 8.39). Except in fire-rated walls, anchor all the studs to the floor and ceiling runner flanges with ½-inch type-S panhead screws.

Cut the crossbracing so that it can be placed between the rows of studs from gypsum board 12 inches high and as wide as the chase wall's width. Space the braces on 48-inch centers vertically and attach to the stud web with screws spaced on maximum 8-inch centers per brace.

Bracing made with 2½-inch (minimum) steel studs can be used in place of gypsum board. Anchor the web at each end of the metal brace to the stud web with two ⅜-inch panhead screws. When the chase-wall studs are not opposite, install steel-stud crossbraces on horizontal 24-inch centers and secure the anchor at the end to form a continuous, horizontal 2½-inch runner. Screw attach the runner to the chase-wall studs within the cavity. Additional chases can be provided in the steel studs by cutting round holes up to ¾ of the stud width, spaced 12 inches apart.

BUILDING SOFFITS

Soffits can be built of gypsum board for either interior or exterior applications. For interior installations, they provide a lightweight, fast, and economical method of filling above cabinets or lockers and housing overhead ducts, lighting systems, pipes, or conduits. Made with steel stud and runner supports, they are faced with screw-attached gypsum board. Braced soffits up to 24 inches deep are constructed without supplemental vertical studs. From Table 8.7, select the components for the desired soffit size. Unbraced soffits without horizontal studs are suitable for soffits up to 24 × 24 inches.

Bracing Soffits

Attach the steel runners to the ceiling and sidewalls as illustrated in Figure 8.40 and place the fasteners close to the outside flange of the runner. On stud walls, space the fasteners to engage the stud. Fasten the vertical gypsum face board to the web of the face-corner runner and the flange of the ceiling runner with type-S screws spaced on 12-inch centers. Place the screws in the face-corner runner at least 1 inch from the edge of the board. Insert steel studs between the face-corner runner and sidewall runner and attach the alternate studs to the runners with metal

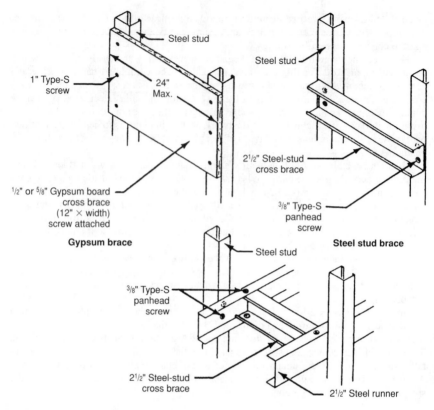

FIGURE 8.39   Three types of chase-wall supports.

lock fasteners or screws. Attach the bottom face boards to the studs and runners with type-S screws spaced on 12-inch centers. Attach the corner bead and finish.

Using Unbraced Soffits

As detailed in Figure 8.41, attach the steel studs to the ceiling and sidewalls by placing the fasteners so that they engage the wall and ceiling framing. Cut the gypsum board to the soffit depth and attach a soffit-length stud with type-S screws spaced on 12-inch centers. Attach the preassembled unit to the ceiling-stud flange with screws spaced on 12-inch centers. Attach the bottom panel with type-S screws spaced on 12-inch centers. Attach the corner bead and finish.

NOTE: Methods for constructing exterior soffits are described in Chapter 6.

ATTACHING FIXTURES

Steel framing provides suitable anchorage for most types of fixtures normally found in residential and commercial construction (Figure 8.42). To ensure satisfactory job perfor-

TABLE 8.7 Braced Soffit Designs (Maximum Dimensions)

Gypsum board thickness[1]	Steel stud size[2]	Maximum vertical[3]	Max. horizontal for max. vertical shown
Inches	Inches	Inches	Inches
½	1⅝	60	48
½	2½	72	36
	3⅝		
⅝	1⅝	60	30
⅝	2½	72	18
	3⅝		

[1]The construction is not designed to support loads other than its own dead weight.

[2]Double-layer applications and ¼-inch board are not recommended for this construction.

[3]Widths shown are based on construction having supplemental vertical studs.

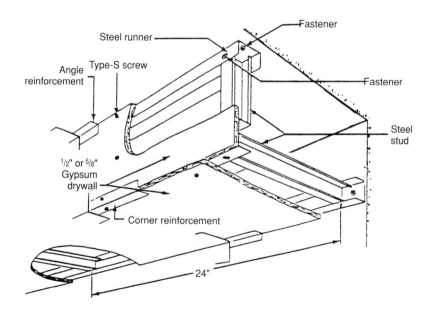

FIGURE 8.40 Typical interior braced soffit.

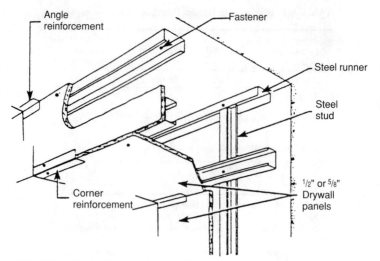

FIGURE 8.41 Typical interior unbraced soffit.

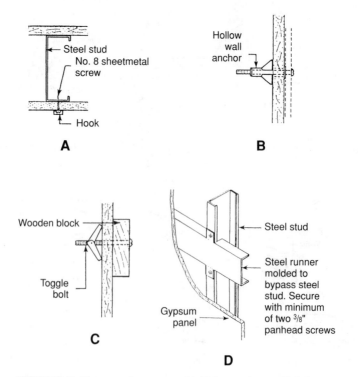

FIGURE 8.42 Fixture attachment types: (A) #8 sheetmetal screw, (B) hollow-wall anchor, (C) toggle bolt, (D) bolt and channel.

mance, it is important to have an understanding of how particular fixtures are attached so that sound-control characteristics are retained and attachments are within the allowable load-carrying capacity of the assembly.

Calculating Loading Capacities

The loading capacities of various fixture attachments with wallboard partitions appear in Table 8.8.

TABLE 8.8 Fixture Attachment Load Data

Fastener type	Size	Base assembly	Allowable withdrawal resistance lbs./ft.	Allowable shear resistance lbs./ft.
Toggle-bolt	⅛		20	40
or hollow-	³⁄₁₆	½" gypsum	30	50
wall anchor	¼	panel	40	60
	⅛	½" gypsum	70	100
	³⁄₁₆	panel and	80	125
	¼	steel stud	155	175
#8 sheet metal screw		½" gypsum panel and steel stud or 25-gauge steel insert	50	80
Type-S buglehead screw		½" gypsum panel and steel stud or 25-gauge steel insert	60	100
Type-S-12 buglehead screw		½" gypsum panel and steel stud or 20-gauge steel insert	85	135
⅜" type-S panhead screw		25-gauge steel	70	120
Bolt welded to 1½" channel	¼		200	250
Two bolts welded to steel insert	³⁄₁₆ ¼	½" gypsum board, plate and steel stud	175	200
		½" gypsum board, plate and steel stud	200	250

TABLE 8.9   Wall Anchor and Toggle-Bolt Capacities

Type of fastener	Size	Allowable load	
		½" wallboard	⅝" wallboard
Hollow-wall screw anchors	⅛" dia. short	50 lbs.	—
	³⁄₁₆" dia. short	65 lbs.	—
	¼", ⁵⁄₁₆" (7.9 mm)	65 lbs.	—
	⅜" dia. short	—	90 lbs.
	³⁄₁₆" dia. long	—	95 lbs.
	¼", ⁵⁄₁₆" (7.9 mm), ⅜" dia. long		
Common toggle bolts	⅛" dia.	30 lbs.	90 lbs.
	³⁄₁₆" dia.	60 lbs.	120 lbs.
	¼", ⁵⁄₁₆" (7.9 mm), ⅜" dia.	80 lbs.	120 lbs.

#8 Sheetmetal Screws. This screw is driven into at least a 25-gauge sheetmetal plate or strip and laminated between the face board and base board in laminated gypsum partitions. The screw can be driven through gypsum board into a steel stud and is ideal for pre-planned light fixture attachments.

Hollow-Wall Anchors. These ¼-inch anchors are installed only in gypsum boards. One advantage to this fastener is that the threaded section remains in the wall when the screw is removed. Widespread spider support is formed when the expanded anchor spreads the load against the wall material, thereby increasing load capacity (Table 8.9).

Toggle Bolts. The ¼-inch toggle bolt is installed only in gypsum board. One disadvantage to the toggle bolt is that the wing fastener on the back falls down into the hollow-wall cavity when the bolt is removed. Another disadvantage is that a large hole is required for the wings to pass through the wall facings.

Continuous Horizontal Bracing. Backup for a fixture attachment is provided with a notched runner attached to steel studs with two ⅜-inch panhead screws.

Installing Shelf-Wall Systems

Load-carrying wall shelves can be installed easily for residential, store, office, or other applications. In steel-framed assemblies, 3¾-inch steel studs on 24-inch centers are secured to the floor and ceiling runners and faced with either single- or double-layer wallboard. The slotted standards are screw-attached to the steel reinforcement, which is inserted between the layers (Figure 8.43). The partition is load-carrying, but with steel studs is not structurally load-bearing. The limiting height is 16 feet.

To install a shelf bracket between the wall studs (Figure 8.44), laminate a 1-×-3-×-12-inch wooden block, or screw-attach the block to the back of a gypsum wallboard panel. The shelf-bracket bearing surface is 1 inch (minimum) of the bracket's width. Hold the bracket in place with 1½-inch #10 wood screws.

Attaching Cabinets

The attachment method shown in Figure 8.45 is appropriate for moderately heavy kitchen and bathroom cabinets and fixtures, except lavatories and wall-mounted toilets. A braced-

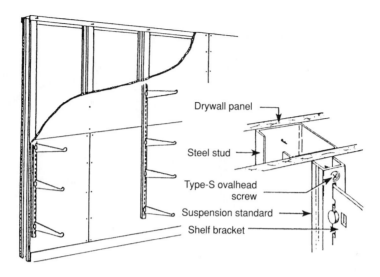

FIGURE 8.43 Typical shelf-wall system.

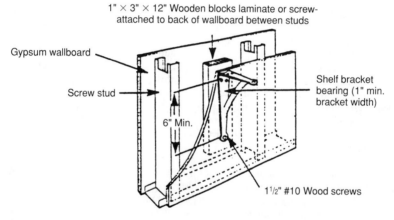

FIGURE 8.44 Typical between-stud installation.

soffit drop ceiling usually is constructed first as described earlier in this chapter. Hang the cabinets from a mounting strip located inside the cabinet. This method is suitable for loads for cabinets weighing up to 67½ pounds for studs or hangers spaced on 16-inch centers and up to 40 pounds for studs or hangers on 24-inch centers.

Hanging Bathroom Fixtures

Regular gypsum wallboard can be used as a base for tile and wall panels in dry areas. Moisture resistant (MR) or water resistant (WR) boards are specially processed gypsum

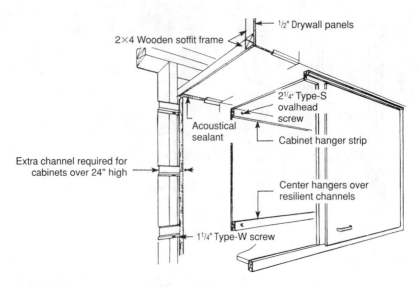

FIGURE 8.45   Typical cabinet attachment.

wallboards that have been designed as a base for ceramic tile and other nonabsorbent fin-ishing materials in wet areas (Figure 8.46). The core, face paper, and back paper of MR or WR board are treated to withstand the effects of moisture and high humidity. A tapered edge usually is provided so that joints can be treated in the normal manner when MR or WR boards extend beyond the tiled area. No special tapes or edge sealants are required. The tile adhesive eliminates the need for further corner treatment, nail spotting, edge seal-ing, or filling in the edge taper in the area to be tiled.

Before the plumber can set the bath fixtures, the interior steel framer must set the fram-ing supports. This includes the supports for the heavy bath fixtures (Figure 8.47). The steel framer also must install all the framing needed to attach the water closet and lavatory car-riers (Figure 8.48).

After the tub, shower pan, or receptor is installed by the plumbing contractor, place tem-porary ¼-inch spacer strips around the lip of the fixture. Precut the panels to the required sizes and make the necessary cutouts. Before installing the panels, brush thinned tile adhe-sive over all cut or exposed panel edges at the utility holes, joints, and intersections.

Install the panels perpendicular to the paper-bound edge that abuts the top of the spacer strip. Fasten the panels with screws on maximum 12-inch centers. If ceramic tile that is more than ⅜ inch thick is used, space the screws on maximum 8-inch centers. When apply-ing MR or WR materials in a bathroom, the framing members must be plumb and true. Place the steel studs not to exceed 24-inch centers. The framing around the tub and shower enclosures should allow sufficient room so that the inside lip of the tub, prefabricated re-ceptor, or hot-mopped subpan is properly aligned with the face of the wallboard. This might necessitate furring out from the studs the thickness of the gypsum board to be used, ½ or ⅜ inch less the thickness of the lip, on each wall that abuts a tub receptor or subpan (Figure 8.49). Frame or block the interior angles to provide a solid backing for the interior corners.

When the framing is spaced on more than 16-inch centers, locate suitable blocking or backing approximately 1 inch above the top of the tub or receptor and at the horizontal joints of the gypsum board in the area that is to receive the tile. Apply the surfacing mate-

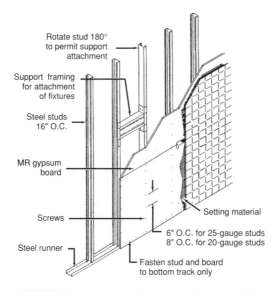

Rotate stud 180°
to permit support
attachment

Support framing
for attachment
of fixtures

Steel studs
16" O.C.

MR gypsum
board

Screws

Steel runner

Setting material

6" O.C. for 25-gauge studs
8" O.C. for 20-gauge studs

Fasten stud and board
to bottom track only

FIGURE 8.46 Details of an MR or WR wallboard installation over steel studs.

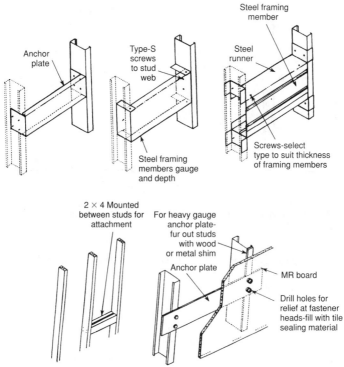

Anchor
plate

Type-S
screws
to stud
web

Steel framing
member

Steel
runner

Steel framing
members gauge
and depth

Screws-select
type to suit thickness
of framing members

2 × 4 Mounted
between studs for
attachment

For heavy gauge
anchor plate-
fur out studs
with wood
or metal shim

Anchor plate

MR board

Drill holes for
relief at fastener
heads-fill with tile
sealing material

FIGURE 8.47 Supports for heavy bathroom fixtures.

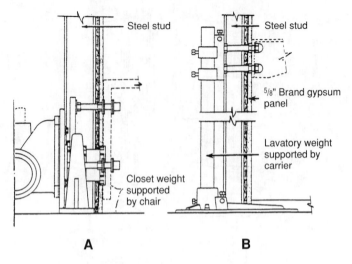

FIGURE 8.48 Typical framing necessary to attach water closet and lavatory carriers.

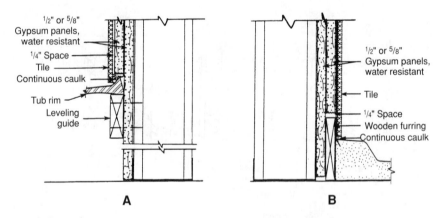

FIGURE 8.49 Fur out from the studs the thickness of gypsum board to be used.

rial down to the top surface or edge of the finished shower floor, return, or tub. Install it so that it overlaps the top lip of the receptor, subpan, or tub and completely covers the following areas (Figure 8.50):

- Over tubs without showerheads, 6 inches above the rim of the tub
- Over tubs with showerheads, a minimum of 5 feet above the rim or 6 inches above the height of the showerhead, whichever is higher
- For shower stalls, a minimum of 6 feet above the shower dam or 6 inches above the showerhead, whichever is higher

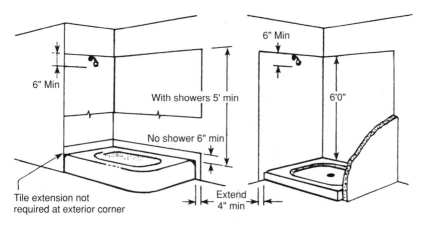

FIGURE 8.50 Bring tile in baths and showers up to the levels shown.

- Beyond the external face of the tub or receptor, a distance of at least 4 inches and to the full specified height
- All wallboard walls with window sills or jambs in shower or tub enclosures, to the full specified height

Make the inside face of the receptor flush with the tiled MR or WR backer board by furring out the studs, if necessary. When the surface finish is a ceramic tile, the spacing of the 2½-inch-thick or less blocking should not exceed 16-inch centers. Space studs 3½ inches or more in thickness on 24-inch centers, provided blocking is used as described previously. Provide appropriate blocking headers to receive soap dishes, grab bars, towel racks, and similar items.

To ensure satisfactory job performance, the fixture attachment should provide sufficient load-carrying capacity within the assembly. Framing and bracing must be capable of supporting the partition elements and fixture additions within L/360 allowable deflection limits. Install the bracing and blocking flush with the face of the framing to keep the stud faces smooth and free of protrusions. Heavy-gauge metal straps mounted on the studs are not recommended for supports because they cause the board to bow and interfere with the flat, smooth application of the MR tiled backer board and ceramic tile.

When heavy plates must be used, fur out the studs with a metal strap or wooden shim to provide an even base for the backer board. If required, the backer board can be ground or drilled to provide relief for projecting bolts and screwheads. Install the shower pans, receptors, tubs, and other plumbing fixtures before erecting the wallboard.

In areas to be tiled, treat all fastener heads with a setting-type joint compound. Fill the tapered edges in the gypsum panel with this setting-type compound, firmly embed joint tape, and wipe off the excess compound. Follow immediately with a second coat over the taping coat. Be careful not to crown the joint. Fold and embed the tape properly in all the interior angles to provide a true angle.

In areas that are not tiled, embed the tape and treat the fasteners with setting-type or lightweight setting-type joint compound applied in the conventional manner. Finish with at least two coats of the setting-type joint compound applied according to directions. Before erecting the tile, seal the cut panel edges of all openings around pipes, fittings, and fixtures with thinned tile adhesive. Remove the spacer strips, but do not caulk the gap at the bottom of the panels.

NOTE: Using an adhesive approved by the tile manufacturer, install tile down to the top edge of the shower floor or tub and overlap the lip or tub return or receptor. Fill all the tile joints with an unbroken application of grout. Apply caulking compound between the tile and shower floor or tub.

Installing Light Fixtures

At light troffers or any ceiling openings that interrupt the carrying or furring channels, install additional crossbracing to restore the lateral stability of the grid (Figure 8.51).

When required for fire-rated construction, use light-fixture protection over the recessed units installed in a direct-screw suspended-ceiling grid. Cut pieces of ½ or ⅝-inch wallboard panels or gypsum-board ceiling panels with type-X board core to form a five-sided enclosure, which is trapezoidal in the cross section (Figure 8.52). Fabricate a box larger than the fixture to provide at least a ½-inch clearance between the box and the fixture, in accordance with local fire regulations.

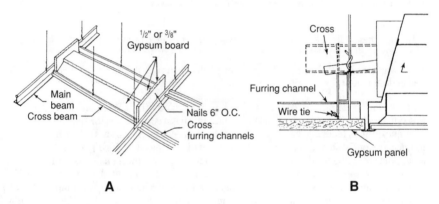

FIGURE 8.51 Typical light-fixture installation.

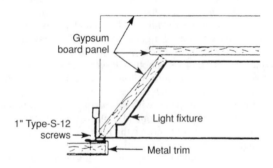

FIGURE 8.52 Typical light-fixture protection.

CHAPTER 9
SPECIAL STEEL-FRAMING APPLICATIONS

Steel framing, as we have seen, offers many advantages. All too often, however, designers and architects fail to take advantage of the fire retardancy, thermal qualities, and sound-control capabilities offered by steel framing. We already have noted that steel framing is nonflammable, but what other fire-rated techniques might be used?

Before considering the actual construction of fire-resistant assemblies, let's take a look at some key terminology and definitions.

Defining Fire Resistance

This term refers to the ability of an assembly to serve as a barrier to fire and to confine the spread of the fire to the area of origin. Fire spreads from one area to another because:

- The barrier collapses
- Openings in the barrier allow the passage of flames or hot gases
- Sufficient heat is conducted through an assembly to exceed specified temperature limitations

These characteristics form the basis for judging when an assembly no longer serves as a barrier in a test.

Rating Fire-Resistance

The fire-resistance rating (FRR) denotes the length of time a given assembly can withstand fire and give protection from it under rigidly controlled laboratory conditions. All tests are conducted under standard conditions. Test methods are described in the American Society for Testing and Materials' (ASTM) *Standard Fire Tests of Building Construction and Ma-*

terials (E119). The standard also is known as *ANSI/UL 263* and *NFPA 251*. The ratings are expressed in hours and apply to floors, ceilings, beams, columns, and walls.

CONSTRUCTING FIRE-RESISTANT GYPSUM ASSEMBLIES

Fire-resistant gypsum drywall is made with a core formulated to offer greater fire resistance than regular wallboard. Generically, these fire-resistant drywall boards are used to prevent the rapid transfer of heat to structural members and to protect them for specified periods of time. The materials are designated as *type-X* products.

ASTM C36 is the standard for gypsum wallboard. According to this standard, type-X gypsum wallboard must provide at least a 1-hour FRR for a ⅝-inch board or a ¾-hour FRR for a ½-inch board applied in a single layer and nailed on each face of the load-bearing wooden framing members. This is the standard when tested in accordance with the requirements of the *Methods of Fire Test of Building Constructions and Materials* or ASTM designation E119. As the number of layers of type-X gypsum increases, the FRR also increases.

An FRR denotes the length of time a given assembly can withstand and provide protection against fire under precisely controlled laboratory conditions. Regardless of the thickness of the installed type-X drywall, the wall still is considered combustible as long as wooden studs are used in the wall. The wall is termed noncombustible only if none of the materials in the partition can burn.

RATING FIRE-RESISTANT WALLS AND CEILINGS

Figure 9.1 shows several FRR wall and ceiling designs that are suitable for all types of steel-framed residential and small commercial buildings, including those with single- and double-layer gypsum-board facings, and other assemblies with sound attenuation materials, fire blankets, and resilient attachments. Here are some typical examples and their ratings.

Installing Fire-Rated Two-Hour Steel-Stud Assemblies

Install the steel-stud framing system (described in Chapter 8), using at least 3½-inch studs and then friction fit 3 inches of insulation. Apply ¾-inch type-X gypsum panel vertically with the wrapped edges parallel to and fully supported by the steel framing. Attach the panels using 1¼-inch-long, type-S or type-S-12 buglehead screws spaced on 8-inch centers along the panel edges and ends, and on 12-inch centers along the intermediate framing.

Installing Fire-Rated Two-Hour Assemblies with C-H-Stud Cavity Walls

Install the framing system using gypsum liner panels. Friction fit a minimum of 3 inches of insulation in the stud cavity. Apply ¾-inch-thick type-X gypsum board vertically with the wrapped edges parallel to and fully supported by the steel framing. Attach the panels using 1¼-inch-long type-S or type-S-12 screws spaced on 8-inch centers along the panel edges and end, and on 12-inch centers along the intermediate framing.

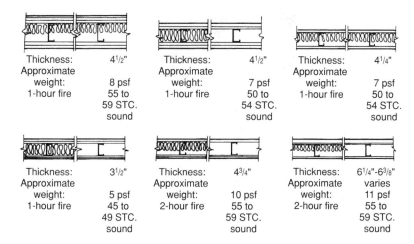

FIGURE 9.1   Fire- and sound-rated steel-framed walls.

Installing Fire-Rated Three-Hour Steel-Stud Assemblies

Install the steel-stud framing and then friction fit the insulation between the studs. Apply a base layer of ¾-inch type-X gypsum board vertically with wrapped edges parallel to and fully supported by the steel framing. Attach the panels to the studs using 1¼-inch-long type-S or type-S-12 drywall screws spaced on 24-inch centers along all the framing. Stagger the board joints by 24 inches on opposite sides of the partition.

Apply the face layer of ¾-inch-thick type-X gypsum panels horizontally with the wrapped edges perpendicular to the steel framing. Stagger the panel end joints 24 inches from the base-layer panel joints. Attach the panels to the steel framing with 2¼-inch-long type-S or type-S-12 screws on 12-inch centers along the framing. Offset the face-layer screws from the base-layer screws by 1 inch. Secure the face panel to the base panel along the horizontal joints of the face layer with 1½-inch-long type-G screws located midway between the steel framing, which is on 24-inch centers, and 1 inch from the horizontal joint.

Stagger the face-panel joints by 24 inches on the opposite side of the partition. Alternatively, the face panels can be installed vertically with the wrapped edges parallel to and fully supported by the framing. Attach the face panels to the steel framing using 2¼-inch type-S-12 screws spaced on 12-inch centers. Stagger the joints in the face and base layer on 24-inch centers.

Installing Fire-Rated, Three-Hour Steel-Stud Chase-Wall Assemblies

Align two parallel rows of floor and ceiling runners spaced 2 inches apart. Attach the rows to the concrete slabs with concrete stub nails or powder-driven anchors on 24-inch centers. On the suspended ceilings, attach with toggle bolts on 16-inch centers, or to wooden framing with suitable fasteners on 24-inch centers. Position the steel studs vertically in the runners, on 24-inch centers. Place the flanges in the same direction and align the studs so that they are directly opposite the studs in the opposite chase.

Cut crossbracing made from gypsum panels 12 inches high by the chase-wall width. Place the bracing between the rows of studs. Space the braces on 48-inch centers vertically and attach to the stud webs with six 1-inch type-S screws per brace. If larger braces are used, space screws on maximum 8-inch centers on each side.

Bracing made from 2½-inch steel studs can be used in place of gypsum panels. Anchor the web at each end of the steel brace to the stud web with two ⅜-inch panhead screws. When the chase-wall studs are not opposite, install the steel stud crossbraces on 24-inch centers horizontally and securely anchor each end to a continuous horizontal 2½-inch runner that is screw-attached to the chase-wall studs within the cavity.

Apply a base layer of ¾-inch, type-X gypsum panels vertically with the wrapped edges parallel to and fully supported by the steel framing. Attach the panels to the studs using 1¼-inch-long type-S or type-S-12 drywall screws spaced on 24-inch centers along all the framing. Stagger the board joints by 24 inches on the opposite sides of the partition. Apply a face layer of ¾-inch type-X gypsum panels horizontally with the wrapped edges perpendicular to the steel framing. Stagger the panel end joints by 24 inches from the base-layer panel joints. Attach the panels to the steel framing using 2¼-inch-long type-S-12 screws on 12-inch centers along all the framing. Offset the face-layer screws from the base-layer screws by 1 inch.

Secure the face panel to the base panel along the horizontal joints of the face layer with 1½-inch-long type-G screws located midway between the steel framing, which is on 24-inch centers, and 1 inch from the horizontal joint. Stagger the face-panel joints by 24 inches on the opposite side of the partition. Alternatively, the face panels can be installed vertically with the wrapped edges parallel to and fully supported by the framing. Attach the face panels to the steel framing using 2¼-inch type-S-12 screws spaced on 12-inch centers. Stagger joints in the face and base layer by 24 inches on center.

Installing Fire-Rated, Four-Hour Steel-Stud Partition Assemblies

Install the steel-stud framing system and then friction fit 2 inches of insulation between the studs. Apply a base layer of ¾-inch type-X gypsum board vertically with the wrapped edges parallel to and fully supported by the steel framing. Attach the panels to the studs using 1¼-inch-long type-S or type-S-12 drywall screws spaced on 24-inch centers along all the framing.

Stagger the board joints 24 inches on the opposite sides of the partition. Apply the face layer of ¾-inch-thick type-X gypsum panels horizontally with the wrapped edges perpendicular to the steel framing. Stagger the panel end joints 24 inches from the base-layer panel joints. Attach the panels to the steel framing using 2¼-inch-long type-S-12 screws on 12-inch centers along all the framing.

Offset the face-layer screws from the base-layer screws by 1 inch. Secure the face panel to the base panel along the horizontal joints of the face layer with 1½-inch-long type-G screws located midway between the steel framing, which is on 24-inch centers, and 1 inch from the horizontal joint. Stagger the face panel joints by 24 inches on the opposite side of the partition. Alternatively, the face panels can be installed vertically with the wrapped edges parallel to and fully supported by the framing. Attach the face panels to the steel framing using 2¼-inch type-S-12 screws spaced on 12-inch centers. Stagger the joints in the face and base layer on 24-inch centers.

INSTALLING TYPE-X PANELS

Apply the type-X panels first to the ceiling at right angles to the framing members and then to the walls. Use boards of maximum practical length so that a minimum number of end joints occur. Bring board edges into contact with each other, but don't force them into place. There is little difference between regular and type-X gypsum board as far as installation methods are concerned.

Attach the steel runners at the floor and ceiling to the structural elements with suitable fasteners located 2 inches from each end and spaced on 24-inch centers. Use toggle bolts or hollow-wall anchors spaced on 16-inch centers to attach the runners to suspended ceilings.

Position the studs vertically, with the open sides facing in the same direction, engage the floor and ceiling runners, and space them on 24-inch centers. When necessary, splice the studs with an 8-inch nested lap and make two positive attachments per the stud flange. Place the studs in direct contact with all the door-frame jambs, abutting partitions, partition corners, and existing construction elements. When studs are installed directly against the exterior walls and the possibility exists that water might penetrate through the walls, install asphalt felt strips between the studs and the wall surfaces.

Anchor all the studs for shelf walls and those adjacent to the door and window frames, partition intersections, corners, and freestanding furring to the ceiling and floor-runner flanges with a metal-lock-fastener tool or screws. Securely anchor the studs to the jamb and the head anchors of the door or borrowed-light frames with bolts or screws. Over the metal door and borrowed-light frames, place a cut-to-length section of runner horizontally with a web-flange bend at each end. Secure the section to the strut studs with two screws in each bent web. Position a cut-to-length stud that extends to the ceiling runner at the vertical panel joints over the door-frame header. When attaching studs to the steel-grid system, determine the grid's structural adequacy to support the end reaction of the wall.

Table 9.1 gives the screw locations and spacing for fire-rated, steel-stud drywall partitions.

When the ceiling size exceeds the following specifications, use control joints. The location of the control joints is the responsibility of the design professional or architect. Isolate the panel surfaces to reduce expansion or contraction with control joints or other means when:

- Partition, furring, or column fireproofing abuts a structural element, except the floor, or dissimilar wall or ceiling

- Ceiling or soffit abuts a structural element, dissimilar wall or partition, or other vertical penetration

- Construction changes within the plane of the partition or ceiling

- Partition or furring run exceeds 30 feet

- Ceiling dimensions exceed 50 feet in either direction with perimeter relief, or 30 feet without relief

- Exterior soffits exceed 30 feet in either direction

- Wings of L-, U-, and T-shaped ceiling areas are joined

- Expansion or control joints occur in the exterior wall

Extend control joints to the ceiling from both corners with less-than-ceiling-height frames. Ceiling-height door frames can be used as control joints. Treat window openings in same manner as doors. Provide an adequate seal or safing insulation (Figure 9.2) behind the control or wall surface when sound or fire ratings are the prime consideration.

Do not break the sheathing behind the control joints. Where vertical and horizontal joints intersect, the vertical joint should be continuous and the horizontal joint should abut it. Caulk splices, terminals, and intersections with a sealant.

Framing at control joints that extend through the wall should have 1½-inch cold-rolled channel (CRC)-alignment stabilizers spaced a maximum of 5 feet vertically. Place the

TABLE 9.1 Screw Locations and Spacing

Face-layer application			Base-layer application			
Screw				Screw		
Length	Type	Spacing and location	Length	Type	Position	Spacing and location
1¼"	S	8" O.C. to studs at joints, 12" O.C. to studs in field	1"			
1⅝"	S	12" O.C. to studs at joints and in field, 12" O.C. to runners	1"	S		12" O.C. to studs at joints and in field, 12" O.C. to runners
1½"	G	Adhesive lamination and supplementary screws	1"	S		8" O.C. to studs at joints, 12" O.C. to studs in field
1⅝"	S	16" O.C. to studs at joints and in field, 12" O.C. to runners	1"	S		16" O.C. to studs at joints and in field
1½"	G	Adhesive strip lamination and supplementary screws	1"	S		12" O.C. to studs at joints and in field
1⅝"	S	12" O.C. to studs at joints and in field	1"	S		12" O.C. to studs at joints and in field
1⅝"	S	8" O.C. to studs at joints, 12" O.C. to studs in field	1"	S		8" O.C. to studs at joints
1⅝"	S	12" O.C. to studs and 24" O.C. to runners	1"	S		24" O.C. to studs at joints and in field
2¼"	S	12" O.C. to studs	1"	S	lst layer	48" O.C. to studs,
2⅝"	S	12" O.C. to studs	1⅝"	S	2nd layer	base layers
1½"	G	between studs at horizontal joints	2¼"	S	3rd layer	12" O.C., face layer
1⅝"	S	12" O.C. to studs	1"	S	lst layer	12" O.C. to studs
2¼"	S	12" O.C. to studs	1⅝"	S	2nd layer	
1½"	G	24" O.C. between studs at horizontal joints				
1"	S	12" O.C. to studs at joints and runners, 8" O.C. to studs in field				
1"	S	8" O.C. to studs at joints 12" O.C. to studs in field				
1"	S	12" O.C. to studs at joints and in field				
2¼"	S	24" O.C. to studs	1¼"	S		24" O.C. to studs at joints and in field
1½"	G	24" O.C. between studs at horizontal joint				

FIGURE 9.2 Cut safing insulation wider than the opening to ensure a compression fit.

channels through the holes in the stud web and securely attach them to the first stud on either side of the control joint.

Control-joint assemblies, suitable for wind pressures up to 40 pounds per square foot (psf), should meet ASTM E514 Class E water-permeance requirements. The backing should have a 6-inch-wide horizontal overlap in the asphalt felt sheathing covering. Place 6-inch-wide asphalt felt strips vertically behind the control joint.

Applying Double-Layers

Double layers consist of a face layer of type-X board over a base layer of gypsum base, which is attached to the framing members.

The standard double-layer, using steel studs and ½-inch type-X board, has an FRR of up to two hours. This rating can be increased by using one of the proprietary or improved type-X boards on the market (Figure 9.3). These FRR assemblies are low cost, much lighter in weight, and thinner than concrete block partitions, and the FRR assemblies offer equivalent performance.

Building Multifamily Structures

All enforced building codes and regulations for single-family structures usually apply to duplexes, apartments, condominiums, townhouses (Figure 9.4), and any structures that house more than one family. In many cases, these regulations are stricter than those for single-family dwellings.

The major construction differences are in the walls between the living units and the floors and ceilings around them. In multifamily buildings, these dividers must be fire-rated. They must be able to slow the spread of fire from one living unit to the next for a minimum period of time. The minimum time might vary from 10 minutes to 1½ hours. An FRR applies not just to the surfacing materials, but to the entire floor, wall, or ceiling system.

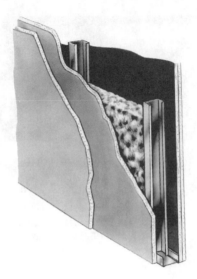

FIGURE 9.3 Double-layer installation onto steel studs with insulation in between.

FIGURE 9.4 Typical multifamily townhouse.

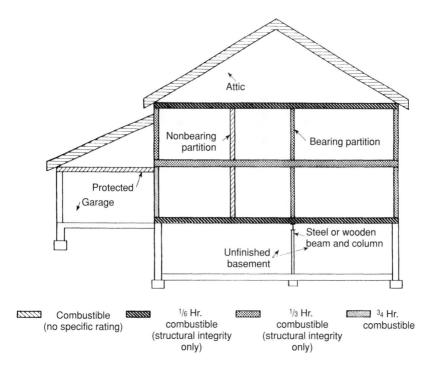

FIGURE 9.5   FRR of a detached single-family house.

Compare the FRR requirement of three buildings: detached single-family house (Figure 9.5), detached two-family house (Figure 9.6), and townhouse (Figure 9.7). In all three buildings, some areas of the structure can be combustible. These areas have no specific fire rating and can burn. Roofs and nonbearing partitions, for example, have no specific rating since they do not affect structural integrity or the soundness of the total construction. In other words, if a roof burns or a nonbearing partition falls down, the remaining structure is not affected.

When a structural system requires a rating such as ⅙ hour for structural integrity only, this means that the structure can burn, but must not collapse from fire for at least ⅙ hour. A ¾-hour combustible rating means that the surfacing material must prevent the fire from attacking the building for a period of at least ¾ hour.

Floor systems, in some buildings, must meet one of three ratings. The floor system between an unfinished basement, which is not used for living space, and first-floor rooms must have a ⅙-hour FRR. Floors between living spaces in the same dwelling unit of a multifamily building must have a ⅓-hour FRR. Floors between different dwelling units must have a 1-hour FRR.

Required ratings for wall systems vary greatly. Generally, in all types of residential buildings, the exterior walls must be designed for a ⅓-hour FRR. Bearing partitions within a dwelling unit also must carry a ⅓-hour rating. Bearing walls between an attached garage and living space must carry a ¾-hour rating. The common wall in a two-family house and some bearing partitions in multifamily buildings must carry a 1-hour rating. Common walls in multifamily buildings must carry a 1½-hour rating. As the danger of loss of life from a fire increases, so does the required time period for fire resistance.

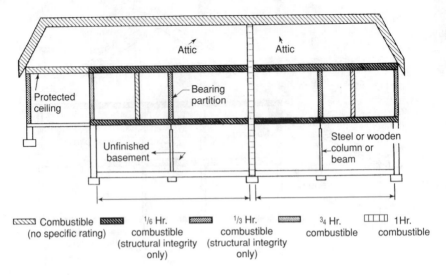

FIGURE 9.6 FRR of a detached two-family house.

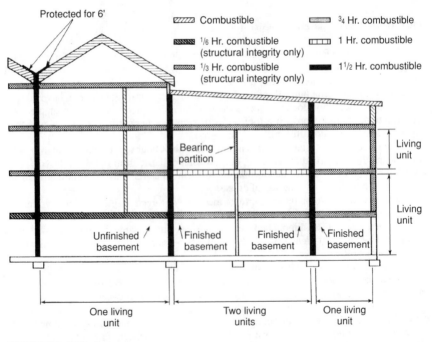

FIGURE 9.7 FRR of a townhouse.

As a general rule, the more fire-resistant or the thicker the finished-wall material applied to the studs, the higher the FRR and STC ratings. Similarly, the use of insulation between studs has a favorable effect. Insulation does not help either structural integrity or sound transmission, but it does improve the FRR.

When an FRR of more than 1 hour is required between living units, standard building practice is to erect a party wall. A party wall is a pair of similar walls with a space between them (Figure 9.8). The two walls bear on the foundation and extend all the way to the roof sheathing. No horizontal structural member passes through the wall. The purpose of the sheathing is to wrap the wall's structural members in a cocoon of fire-resistant material. Then if a fire starts in one living unit, it cannot easily spread to adjacent living units.

The general steel-framing construction procedure for building a party wall is similar to that for ordinary walls, except where the closeness of the two walls to each other forces changes.

Building Area-Separation Walls

Figure 9.9 illustrates, in dramatic fashion, the value of breakaway area-separation walls. The fire destroyed one townhouse unit, but the area-separation wall arrangement protected

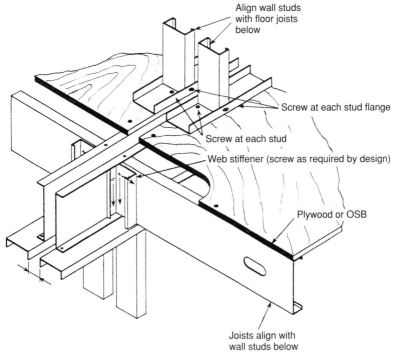

NOTE: In order to facilitate the attachment of drywall which must extend to the underside of floor sheathing, blocking must be provided between each joist.

FIGURE 9.8   Typical party wall.

FIGURE 9.9 The fire protection of the gypsum board area-separation wall system is demonstrated in dramatic fashion by the results of this actual townhouse fire. The 2-hour fire-rated assembly performed as expected and protected the adjacent properties. The breakaway feature allowed the fire-side structural framing to collapse without pulling down the entire wall.

the two adjacent properties. In addition to providing 2-hour fire protection, area-separation walls also are good sound barriers between units in multifamily wood-framed buildings. Some codes identify this type of wall by other names, such as fire wall or barrier, lot-line wall, or party wall. These systems can be used in buildings up to four stories high and offer up to 3-hour FRRs and up to an STC 57. As can be seen in the illustrated construction, these systems are a good example of how wooden and steel framing can work together.

There are two basic area-separation wall systems: solid and cavity. Each system can be erected quickly by carpenters and/or drywall tradespeople. Both systems are continuous, vertical, nonload-bearing wall assemblies of gypsum panels attached to steel framing. They extend from the foundation up to or through the roof and act as barriers to fire and sound transmission. These walls are erected one floor at a time, beginning at the foundation and continuing up to or through the roof. At the intermediate floors, install metal floor-ceiling track back to back to secure the top of the lower section of the partition and the bottom of the next section to be installed.

At intermediate floors and other specified locations, attach the walls to the adjacent framing on each side with aluminum clips that soften when exposed to fire. These so-called *breakaway* clips, described in Chapter 2, are wide and have legs that are either 1 × 2 inches, 1 × 2½ inches, or 2 × 2 inches. Attach the clips to the wooden framing at each vertical steel-framing stud or wood-framed intersection using one ⅜-inch type-S panhead screw to secure it to the steel and one 8d nail or one 1¼-inch type-W drywall screw to secure it to the wooden framing. The recommended location of these clips is on the lower side of the structural wooden framing. In that location, the clip provides the greatest assurance that the area separation wall will remain in place and be structurally sound should one of the adjacent structures fail.

Solid Separation Walls. This system is designed for use between load-bearing walls. While the total assembly is thicker than cavity walls, solid walls are no thicker than masonry barriers when used in the same type of construction.

Foundation Installations. Position the 2-inch-wide steel C runner at the floor and securely attach it at both ends to the foundation with powder-driven fasteners spaced on 24-inch centers (Figure 9.10). Space adjacent runner sections ¼ inch apart. When specified, caulk the runner at the foundation with at least a ¼-inch bead of acoustical sealant.

First-Floor Installations. Install the panels and steel H studs cut to a length that is more than the floor-to-floor height. Install two thicknesses of 1-inch liner panels vertically in the C runner with the long edges in the H stud. Erect the H studs and double-thickness liner panels alternately until the wall is completed. Cap the ends of the run with the vertical C runner and fasten it to the horizontal C runner flange with ⅜-inch type-S screws.

Intermediate-Floor Installations. Cap the top of the panels and studs with back-to-back C runners, that are screwed together with double ⅜-inch type-S screws at the ends and spaced on 24-inch centers. Fasten the steel studs to the runner flange on alternate sides with ⅜-inch screws. Secure the studs to the framing with aluminum breakaway angle clips that are screw-attached to both sides of each stud and the framing (Figure 9.11). Except at the foundation, install fire blocking between the joists and the fire barrier.

Roof Installations. Continue erecting the studs and panels for succeeding stories as previously described. At the roof, cap the panels with a C runner and fasten it to the studs with ⅜-inch screws (Figure 9.12). Fasten the studs to the framing with aluminum breakaway clips.

To assure a fire-rated assembly that does not need battens, maintain a minimum ¾-inch air space between the steel H stud assembly and any adjacent framing members. When a ¾-inch air space cannot be maintained, cover the steel H stud and H-stud tracks with screw-attached 6-inch-wide battens fabricated from ½-inch type-X gypsum wallboard. Alternately, fasten the ½-inch type-X gypsum wallboard panels to the H studs and cover

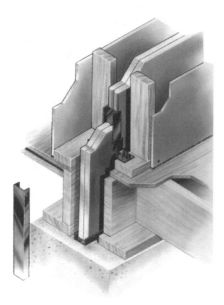

FIGURE 9.10 Foundation of a solid separation wall.

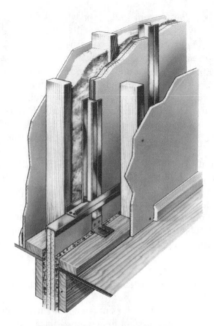

FIGURE 9.11 Intermediate floor of a solid separation wall.

FIGURE 9.12 Roof details of a solid separation wall.

the joints with tape-and-joint compound to provide a finished wall. Install mineral wool or glass fiber in adjacent cavity-shaft walls to provide higher STC ratings.

Cavity-Separation Walls. Cavity-separation walls are intended for use in buildings where load-bearing walls are not used at the line of separation between units. For that reason they are the thinnest of all fire walls. For instance, a typical 3-hour area separation wall is a mere 4⅞ inches thick. The 2-hour version is only 3½ inches thick.

Foundation Installations. Position the 2½-inch steel C runner on the floor and attach it at both ends to the foundation with powder-driven fasteners spaced on 24-inch centers (Figure 9.13). Caulk the runner on the floor with a minimum ¼-inch bead of acoustical sealant.

First-Floor Installations. Cut the 1-inch gypsum shaft-liner panels and steel studs to convenient lengths that are more than floor-to-floor height. Erect the liner panels vertically in the C runner with the long edges in the groove of the C-H stud. Install the steel C-H studs between the panels and cap the ends of the run with an E stud or C runner. Fasten the studs to the bottom of the C runner on alternate sides with ⅜-inch type-S screws.

Intermediate-Floor Installations. Cap the top of the panels and studs with a steel C runner and fasten it to the C-runner flanges on the alternate sides with ⅜-inch type-S screws. Install the bottom C runner for the next row of panels over the top runner and stagger the end joints by at least 12 inches (Figure 9.14). Fasten the runners together with double ⅜-inch screws at the ends and spaced on 24-inch centers. Secure each steel C-H stud to the framing with the aluminum breakaway clip (Figure 9.15). Fasten the clip to both sides of each stud with ⅜-inch screws and to the framing or subflooring with 1¼-inch type-W screws.

Attachments can vary for convenience, but are required within 6 inches of each end of the stud and spaced on not more than 10-foot centers. The overall wall height must not exceed 44 feet. Except at the foundation, install fire blocking between the joists and the fire barrier.

NOTE: Use 4-inch 20-gauge steel C-H studs on floors below the top 23 feet of the building.

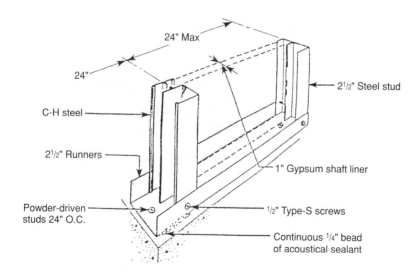

FIGURE 9.13 Foundation of a cavity-wall separation system.

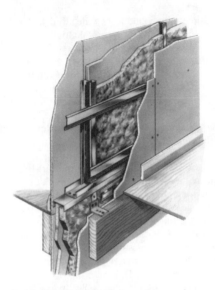

FIGURE 9.14 Intermediate floor of a cavity-wall separation system.

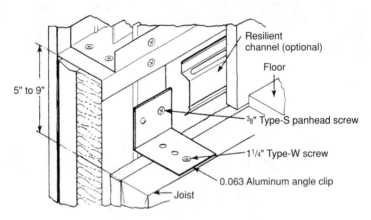

FIGURE 9.15 Installing the breakway clip and the resilient channel in the intermediate floor-cavity wall.

When required by the design, install resilient channels horizontally to the face side of the studs, 6 inches above the floor and 6 inches below the ceiling joists on maximum 24-inch centers. Attach the channels to the studs with ⅜-inch type-S screws driven through the holes in the mounting flanges. Extend the channels to the ends of the runs and attach them to the C runners. Splice the channels by nesting them directly over the studs and then screw-attach them through both the flanges. Reinforce with screws at both ends of the splice.

Roof Installations. Continue erecting the studs and panels for succeeding stories as previously described. At the roof, cap the panels with a steel runner. Fasten the stud to the framing with aluminum clips within 6 inches of the roofline.

INSTALLING SOUND-CONTROL ASSEMBLIES

Structural assemblies, as we have seen, are fire-rated for the safety of the building's occupants. Assemblies also are sound-rated for the comfort and privacy of occupants in multifamily dwellings. As the steel framer, it is important that you understand sound control. Noise is unwanted sound. Music generated by a stereo might be pleasing to the person playing it, but very disturbing to his or her neighbor.

The architects or building designers take the first step in sound control by following the principles of good-room arrangement in the floor plan. They take the second step by specifying structural systems and finishing materials that control sound. Builders or contractors take the next step by rigidly following plans and specifications. Finally, the steel framer completes the task by properly installing the frame.

Sound is produced by a vibrating object. Sound travels from the source in all directions. It is a pressure wave in the air and travels the same way a ripple moves away from a pebble thrown into a lake. The greater the intensity of the vibration, which is called *amplitude*, *pressure*, or *magnitude*, the louder the sound. Sound also has frequency, or *pitch*, as in a high-pitched squeal or a low-pitched rumble.

Sound control is the art of putting something between a person's ears and the source of the sound to reduce its intensity, or doing something to reduce the sound at its source. If the ears and the sound source are in the same room, the room can be treated to reduce the amount of sound that bounces off the walls, floors, and ceiling. Sound control is not soundproofing. If something is waterproof, no water can get into it. Similarly, if a room or a structure is soundproof, absolutely no unwanted sound can penetrate into it or escape from it. That degree of sound isolation is neither practical nor affordable for everyday living spaces. In fact, it is not even desirable because it cuts off all sense of contact with the outside world.

There are three sources of unwanted noise in any given area:

- Noise from outside sources, such as traffic, children at play, neighbors, barking dogs, aircraft, and the like
- Noise originating in another part of the building and transmitted through walls, floors, and ceiling
- Noise generated inside the same area

Most sound travels through both air and solid objects. For example, the sound of one's voice is airborne sound until it strikes a wall and becomes structure-borne. The wall vibrates and reradiates the sound as airborne sound in an adjacent room or area, but at a lower level because some of the sound intensity has dissipated inside the wall. At each physical barrier, the cycle repeats until the sound dissipates completely. There are two types of sound transmission.

Airborne Sounds. As the name implies, airborne sound is carried through the air. Conversations, the electronic output from televisions, stereos, and radios, as well as the hum of motors are typical examples (Table 9.2). Airborne noise can travel over, under, or around walls, through the windows and doors adjacent to them, through air ducts, and through the floors and crawl spaces below. These flanking paths must be correctly treated

TABLE 9-2 Sound Transmission Level Recommendations (from HUD Minimum Property Standards 1- and 2-Family and Multifamily Dwellings)

Location of partition	STC	
Separating living unit from other living units, common service area or public spaces (average noise)[1]	45	
Separating living units from other living units, public spaces or common service areas (high noise)[2]	50	

Location of floor-ceiling	STC	ICC
Separating living units from other living units, public spaces or common service areas[1,3]	45	45
Separating living units from public spaces or common service areas (high noise)[2]	50	50

[1]Public spaces of average noise include entries, stairways, etc. Assume floors in corridors are carpeted; otherwise increase STC by 5.

[2]Areas of high noise include boiler rooms, mechanical-equipment rooms, central laundries and most commercial uses.

[3]Does not apply to floor above storage rooms where noise from living units would not be objectionable.

to reduce the transmission of sound. Hairline cracks and small holes increase the transmission of sound.

Structure-Borne Sound. Structure-borne sound, also called *impact sound*, is sound carried by the building's structure. The thud of shoes being dropped, the vibration of some kitchen and laundry appliances, and plumbing noise are examples are structure-borne sounds. When measured, the rating is called an *impact insulation class* (IIC). These values rate the capacity of floor assemblies to control impact noises, such as footfalls. Minimum values established by the U.S. Department of Housing and Urban Development (HUD) range from IIC 45 to 50 (Table 9.2).

Controlling Sounds

The term *sound control* refers to the ability to attenuate sound passing through a partition. The *sound transmission class* (STC) commonly is used as a rating of sound attenuation performance. It is accurate for the sounds of speech, but not for music, mechanical-equipment noise, or any sound with substantial low-frequency energy. STCs are tested per ASTM E90 and rated per ASTM E413.

Each type of construction has an STC rating. The STC is a measure of the reduction of the transmission of airborne sound. The higher the rating, the more soundproof the construction. The significance of STC numbers is illustrated in Table 9.3 from the Acoustical and Insulation Materials Association (AIMA).

The *impact insulation class* (IIC) is a numerical evaluation of a floor-ceiling assembly's effectiveness in retarding the transmission of impact sound. Determined from laboratory testing, the testing standard is ASTM E492 and the rating is given per ASTM E989.

The *noise reduction coefficient* (NRC) is a measure of sound absorption. This is an important consideration for controlling acoustics within a confined area. It generally does not apply to the performance of a structural system.

TABLE 9-3 Sound Transmission Control (STC) Ratings

Rating	
25	Normal speech can be understood quite clearly.
30	Loud speech can be understood fairly well.
35	Loud speech audible but not *intelligible*.
42	Loud speech audible as a murmur.
45	Must strain to hear loud speech.
48	Some loud speech barely audible.
50	Loud speech not audible.

NOTE: When comparing rated constructions, remember that 3 decibels (dB) is the smallest difference that the human ear can detect clearly. Thus differences of 1 or 2 points can be considered negligible. Also note that even this general comparison is valid only with respect to the given level of background noise.

Applying Sealant

If a partition is to effectively reduce the transmission of sound, it must be airtight at all points. The perimeter must be sealed with an acoustical sealant, which is a caulking material that remains resilient and does not shrink or crack.

To properly seal a partition perimeter, cut the wallboard for a loose fit around the partition perimeter (Figure 9.16). Leave a groove no more than ⅛ inch wide. Apply a ¼-inch round bead of sealant to each side of the runners, including those used at the partition intersections with dissimilar wall construction. Immediately install the boards and squeeze the sealant to achieve firm contact with the adjacent surfaces. Fasten the boards in the normal manner. Drywall panels can have the joint treatment applied in a normal manner over sealed joints or the panels can be finished with base or trim as desired.

FIGURE 9.16 Applying acoustic sealant.

To be effective, the sealant must be properly placed. Placement is as important as the amount used. The following are some of the important locations where acoustical sealant should be placed.

Control Joints. Apply sealant behind the control joint to reduce the path of sound transmission through the joint.

Partition Intersections. Seal the intersections with sound-isolating partitions that are extended to reduce sound flanking paths.

Openings. Seal openings by applying the sealant around all cutouts, such as those for electrical boxes, plumbing, medicine cabinets, heating ducts, and cold-air returns. Caulk the sides and backs of electrical boxes to seal them. Do not install any fixtures back to back.

Door Frames. Apply a bead of sealant in the door frame just before inserting the face panel.

Provide perimeter relief and reduce the likelihood of sound flanking around related construction for gypsum construction surfaces where:

- The partition or furring abuts a structural element, except the floor, dissimilar walls, or ceiling materials
- The ceiling abuts a structural element, dissimilar partition, or other vertical penetration
- Ceilings with dimensions that exceed 30 feet in either direction at columns

Isolation is important to reduce potential joint problems in partitions, ceilings, and beam and column furring (Figure 9.17). Generally, the following procedures for isolating surfaces remedy any problems.

Wallboard Edge Treatments. Where boards intersect dissimilar materials or structural elements, apply appropriate trim to the face-layer perimeter and apply acoustical sealant to close the gap. Some acoustical vinyl trim can be used without sealant or joint treatment.

Partitions-Structural Ceilings. Attach the steel runner to the structural ceiling to position the partition. Cut the steel stud ⅜ inch less than the floor-to-ceiling height. Attach the gypsum to the stud at least 1½ inches below the ceiling. Allow a ⅜-inch minimum clearance atop the gypsum boards and finish as required.

Partition-Exterior Walls or Columns. Attach the steel stud to the exterior wall or column to position the partition. Attach the gypsum board only to the second steel stud

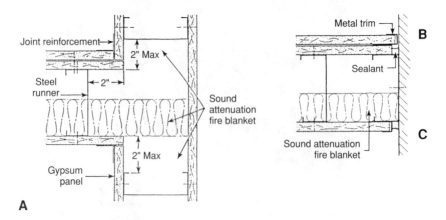

FIGURE 9.17 Use acoustical sealant at (A) the sound-isolating partition intersection, (B) partition-wall intersection, and (C) partition control joint.

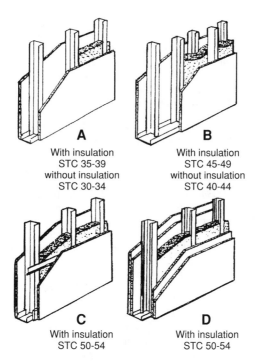

A
With insulation
STC 35-39
without insulation
STC 30-34

B
With insulation
STC 45-49
without insulation
STC 40-44

C
With insulation
STC 50-54

D
With insulation
STC 50-54

FIGURE 9.18 Sound ratings with and without insulation: (A) Adding a sound absorbing insulation increases the STC of the wall by four points; (B) Staggered studs provide partially separated framing for wall surfaces and increase the STC about eight points more than the wall in A; (C) Mounting resilient gypsum surfaces raises the STC about 12 points above that of the wall in A; (D) Two-ply construction increases the weight of the wall surfaces and helps provide the wall with a higher obtainable sound resistance rating of STC 50 to 54.

erected vertically at a maximum of 6 inches from the wall. Allow at least a ¾-inch clearance between the partition panel and the wall. Caulk as required with acoustical sealant. Sounding ratings can be increased as shown in Figure 9.18 by the following means.

- Add a sound absorbing insulation to increase the wall's STC by four points.
- Stagger studs to provide partially separated framing for wall surfaces and increase the STC by about eight points from the wall in A.
- Resilient mounted wallboard surfaces raise the STC by about 12 points above that of the wall in A.
- Two-ply construction increases the weight of the wall surfaces and helps provide the wall with a higher obtainable sound resistance from STC 50 to 54.

Fixture Locations. Attaching fixtures to sound-barrier partitions might impair the sound-control characteristics. Only attach lightweight fixtures to resilient wall surfaces that are constructed with resilient channel, unless special framing is provided. Refrain from attaching fixtures to party walls that provide a direct-flow path for sound. Gypsum boards used in the ceiling are not designed to support light fixtures, troffers, air vents, or other equipment. Separate supports must be provided.

Installing Resilient Furring Channels

One of the ways to achieve sound-isolation control is to install resilient furring. Whether the furring is installed by the framing group or the drywall installers, both contractors must be familiar with its installation.

Resilient furring channels are fabricated of galvanized steel with expanded metal legs that provide resiliency to reduce sound transmission through steel-wall and ceiling-framing assemblies (Figure 9.19). The steel channels float the finished board away from the studs and joists and provide a spring action that isolates the gypsum board or material from the framing. In other words, these channels act as shock absorbers by reducing the passage of sound through the wall or ceiling and thus increase the STC rating. In addition, this spring action also tends to level the panel surface when it is installed over uneven framing.

When applying resilient channels to steel framing, attach the steel runners on the floor and ceiling to the structural elements with suitable fasteners located 2 inches from each end and spaced on 24-inch centers. Position the studs vertically, face the open sides in the same direction, engage the floor and ceiling runners, and space them on 24-inch centers. Anchor the studs to the floor and ceiling runners on the resilient side of the partition. Fasten the studs to the runner flange with metal lock fasteners or ⅜-inch type-S panhead screws.

Position the resilient channel at right angles to the steel studs, spaced on 24-inch centers, and attach them to the stud flanges with ⅜-inch type-S panhead screws driven through the holes in the channel-mounting flange. Install the channels with the mounting flange down, except at the floor, to accommodate the attachment. Locate the channels 2 inches from the floor and within 6 inches of the ceiling. Splice the channel by nesting it directly over the stud and screw-attach it through both flanges. Reinforce with screws located at both ends of the splice (Figure 9.20).

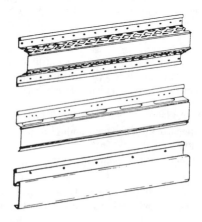

FIGURE 9.19 Typical resilient furring channels.

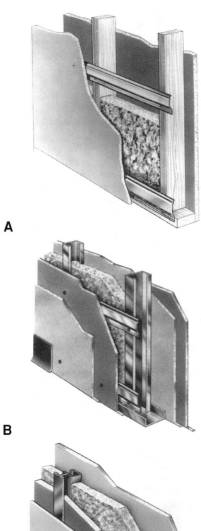

A

B

FIGURE 9.20 Installation methods for various resilient channel applications: (A) single-layer application, (B) double-layer, high-performance wall, (C) cavity-shaft wall installation.

C

Framing Separating Walls

Walls transmit sound vibrations from one face to another through the studs. Vibrations through the joists transmit sounds between ceilings and floors. Eliminate framing members common to both sides and break the path. In walls, stagger the studs or build two walls with a gap in the center. A staggered steel-stud arrangement is shown in Figure 9.21.

If a second wall is built, construct it on one side and place it close to but not touching the existing wall. It also helps to add insulation batts between the new studs. These batts can touch the existing wall. Installing a layer of sound-deadening board beneath the new wall's gypsum board also is beneficial.

Heavy materials reduce sound better than light materials. For example, placing a sound-deadening board under the gypsum board and applying a second layer of gypsum board, or even adding wooden or hardboard paneling to a wall, increases the STC rating.

Sound-deadening backer board comes in 4-×-8-foot sheets that can be screwed to studs or joists. There are two kinds of sound-deadening board. One is made of organic fibers and the other is a form of gypsum board. Ordinary gypsum board is applied over the sound-deadening board to provide a finishing surface. Attach the gypsum board to the sound-deadening underlayer only with adhesive so there is no hard connection to the studs or joists.

Applying Insulation Blankets

Many partition systems have been developed to meet the demand for increased privacy between units in residential and commercial construction. Designed for steel-stud construction, these assemblies offer highly efficient sound-control properties, yet they are more economical than other partitions that offer equal sound isolation. These improved sound-isolation properties and ratings are obtained by using sound-attenuated fire-insulation

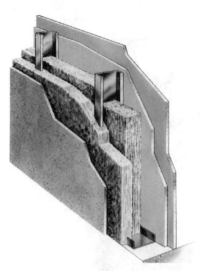

FIGURE 9.21 Method of staggering steel studs.

blankets and decoupling the partition faces. Decoupling is achieved with a resilient application or with double rows of studs on separate plates.

It is possible to erect insulation vertically and hold it in place with Z-furring channels spaced on 24-inch centers. Except at the exterior corners, attach the narrow flanges of the furring to the wall with concrete stub nails or powder-driven fasteners spaced on 24-inch centers. At the exterior corners, attach the wide flange of the furring channels to the wall and extend the short flange beyond the corner. On the attached-wall surface, screw-attach the short flange of the furring channel to the web of the attached channel.

Start from this furring channel with a standard-width insulation panel and continue in the regular manner. At the interior corners, space around the second channel no more than 12 inches from the corner and cut the insulation to fit. Hold the insulation in place until the gypsum wallboard panels are installed with 10-inch-long staples that have been field-fabricated from an 18-gauge tie-wire and inserted through the slot in the channel. Apply wooden blocking around the window and door openings and as required to attach fixtures and furnishing.

Apply the gypsum panels parallel to the channels. Place the vertical joints over the channels. Do not use end joints in single-layer application. Attach the gypsum panels with 1-inch type-S screws spaced on 16-inch centers in the panel field and at the edges. Use 1¼-inch type-S screws spaced on 12-inch centers at the exterior corners. For double-layer applications, apply the base layer parallel to the channels and the face layer either perpendicular or parallel to the channels with the vertical joints offset by at least one channel. Attach the base layer with screws on 24-inch centers and the face layer with 1⅝-inch screws on 16-inch centers.

Installing Sound-Attenuating Fire Blankets

Install these blankets in the stud cavities of sound-rated partitions. They can be held in place with a friction fit (Figure 9.22) or by perforated web tabs that are built into the steel framing. With either method, the insulation also can be anchored in place with specialized steel studs. To fit the blankets in the cavity, it is sometimes necessary to make a slit-cut down the center and partially through the blanket (Figure 9.23). This allows the blanket to flex or bow in the center, eases the pressure against the studs, and transfers the pressure to the face panel. This dampens the sound vibrations more effectively. Screw-attach the panels.

Aluminum-foil-faced blankets are similar to paper-enclosed regular blankets, but have a highly reflective aluminum foil laminated to the vapor-retarder side to protect against condensation. A minimum adjoining air space of ½ inch in the sidewalls and 1 inch in the ceilings is required to obtain an insulating value from the foil reflectivity. This blanket can be used in ceilings, walls, and floors with air space. It is most effective with air conditioning and in areas of extreme summer temperatures.

TAKING ADVANTAGE OF THERMAL INSULATION

The resistance to heat flow through a wall is usually qualified by a property known as the *thermal* resistance or *R-value*. Research has shown that a wall's R-value is not affected by the thickness of the steel stud. Because the stud-web thicknesses are small compared to other dimensions, the heat conducted through them is somewhat limited. When higher R-values are required, the use of an exterior insulative sheathing such as extruded polystyrene or polyisocianurate foam forms an effective thermal break and significantly increases the R-value.

FIGURE 9.22   Installing sound insulation between steel studs.

FIGURE 9.23   If insulation is wider than the space between the studs, cut a slit-field down the center and partially through the blanket to allow it to flex or bow in the center and thus fit.

When these sheathings are used, calculations using the American Society of Heating Refrigerating and Air Conditioning Engineers' (ASHRAE) zone method tend to underestimate the R-value unless the newly developed zone-correction factors are used.

In general, a stud in a wall acts as a *thermal bridge*. It provides a path that conducts heat more rapidly than a path located midway between studs. Both wooden and steel studs provide thermal bridges as illustrated in Figure 9.24. Part of the heat flows directly through the cavity, while part flows laterally toward the stud and then combines with that flowing directly through the stud. Either type of wall can be designed to provide the desired thermal performance. Figure 9.25 shows a map of the 10 insulation zones in North America. Table 9.4 gives the recommended total R-values for existing residences in the North American zones.

TABLE 9.4 Zone Recommended R-Values

	Ceilings below ventilated attics	Floors over unheated crawl spaces	Basements	Exterior walls/ crawl-space walls
1	R-19	0	0	R-11
2	R-19	0	0	R-11
3	R-30	0	R-11	R-11
4	R-30	R-11	R-11	R-19
5	R-30	R-11	R-11	R-19
6	R-38	R-19	R-11	R-19
7	R-38	R-19	R-11	R-19
8	R-38	R-30	R-19	R-30
9	R-49	R-30	R-19	R-30
10	R-49	R-38	R-30	R-38

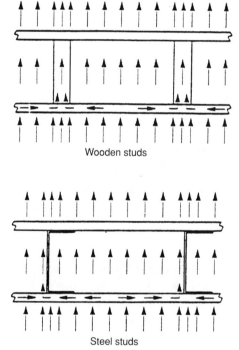

Wooden studs

Steel studs

FIGURE 9.24 Heat flow through steel and wooden framing.

Studs, joists, and ceiling areas can accept blown or semirigid board insulation as thermal needs dictate. Most of the insulation used with steel framing are either blankets or batts made of mineral wool or fiberglass. Install the blankets to completely fill the height of the stud cavity and with the vapor retarder facing inside or outside of the wall according to the

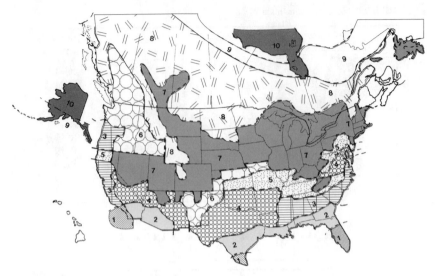

FIGURE 9.25   Recommended total R-values.

job specifications. To tightly fill the height, if necessary, cut the stock-length blankets with a serrated knife before inserting them in the void. Tightly butt the ends and sides of the blankets within the cavity. Cut small pieces of the blankets for narrow stud spaces next to door openings or at partition intersections. It also is recommended that the full-width batt insulation be placed in all jambs, headers, and built-up members, as well as between framing members in areas that separate climate-controlled spaces from the exterior. The use of a thermal break or rigid insulation on exterior framing also can result in significantly better thermal performance. Other insulation tips were given earlier in this chapter.

In ceilings, fit the insulation around recessed lighting fixtures. Do not cover fixtures with insulation because heat can build up and cause a fire. Insulate the following areas in a steel-framed residence:

- Between all studs in exterior walls
- Between structural members and door and window frames on all four sides
- In small spaces between window- and door-framing members
- Behind outlet boxes and entrance panels in exterior walls
- Against second-floor header joists and edge joists
- Between the top plate and subfloor of two-story and split-level houses
- Between ceiling joists of unheated garages or porches when the rooms above are used for living purposes
- Between ceiling joists below unheated attics
- Below stairways to unheated attic space
- In knee walls of heated attics
- Between or over joists of decks above heated living space
- Over ceiling joists of rooms with sloping ceilings

- On basement walls when spaces are used for living purposes
- Against header and edge joists between the sill and subfloor of houses with basements
- Between floor joists over a crawl space
- Around the inside perimeter of concrete-slab floors in heated rooms

Wherever insulation is used, a vapor retarder is required. The proper use and placement of vapor retarders are important factors in modern, energy-efficient construction (Figure 9.26). Improper placement of a vapor retarder can produce condensation in the exterior-wall stud cavities and cause the structure to deteriorate.

The location and placement of the vapor retarders should be determined by a qualified mechanical engineer. For example, two vapor retarders on opposite sides of a single wall can trap water vapor between them and create moisture-related problems in the core materials.

When a polyethylene vapor-retarder film is installed on ceilings behind gypsum panels under cold conditions, install the ceiling insulation, whether batts or blankets, before installing the boards. If loose-fill insulation is to be used above the ceiling, it must be installed immediately after the ceiling board is installed during periods of cold weather. The plenum or attic space also should be vented properly. Failure to follow this procedure can result in moisture condensation behind the wallboard panels, which might cause them to sag.

FIGURE 9.26 Installation of vapor barrier.

INSTALLING CURTAIN-WALL INSULATION

As described in Chapter 6, exterior walls are nonaxial-load-bearing exterior walls. Curtain-wall insulation provides fire resistance, sound isolation, and thermal performance for exterior curtain walls. Install the insulation batts behind spandrel panels of glass, concrete, marble, or granite for thermal control and to ensure spandrel integrity. Proper spandrel panel protection helps eliminate the leapfrog effect by keeping fire from lapping around the spandrel panels and reentering the building at higher floors.

Curtain-wall insulation usually is made of mineral fiber and, when used in conjunction with a safing insulation, offers fire protection to 2000 degrees F. The safing insulation is installed between the floor-slab perimeter and the curtain-wall assembly and should fill all voids. Proper installation, especially with support brackets or impaling clips, eliminates the chimney effect and stops the passage of fire upward around the floor-slab perimeter (Figure 9.27).

INSPECTING THE JOB

Proper job inspection begins when the steel frame is being constructed and continues immediately after its completion. Inspecting work during these periods can eliminate potential problem areas or procedures that result in unsatisfactory work. Immediate corrective actions usually are less costly than callbacks to repair and perhaps rebuild walls and ceilings after the job is completed.

A complete understanding of job details, schedules, and specifications is necessary to conduct a proper inspection. If the assembly is to meet fire- and sound-rating requirements, then construction details must be understood and followed. All walls and ceilings

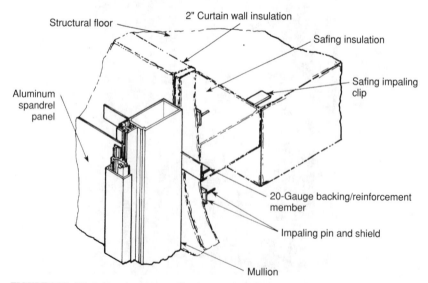

FIGURE 9.27 Eliminating the chimney effect in a curtain wall.

must be judged by these criteria and the contract conditions. It is important that drawings and specifications be complete, accurate, and easily understood.

The inspection phase is most important and, in many cases, determines the success of the job. Make an accurate check of the following major categories to obtain the best results.

Schedule Framing Inspections. Conduct framing inspections at the following stages.

- When the job is almost ready for materials deliveries, check environmental conditions and delivery-acceptance plans.
- When materials are delivered to the job.
- When the framing is erected, but before board or lath is applied.
- When the job is completed.

Framing Inspections. Steel-framing members must meet the architect's or designer's specifications and be free of defects. During and after the framing construction, make the following inspections.

- Ensure that the steel-framing materials meet the specifications required by local building codes, regulations, and standards.
- Check the accuracy of alignment and the position of the framing members, including required bracing, against the plans and details.
- Make sure load-bearing studs are directly underneath the members they support.
- See that the partitions are straight and true and that the ceilings are level.
- Measure the spacing of studs and joists. Spacing should not exceed the maximum allowable for the system.
- Look for protrusions of blocking, bridging, or piping, and twisted studs and joists that create an uneven surface. Correct the situation before attaching boards.
- Make sure there is appropriate blocking and support for fixtures and boards.
- See that window and door frames and electrical and plumbing fixtures are set for the board thickness used.
- Check for the proper position and attachment of resilient and furring channels.
- Review all steel framing for compliance with the minimum framing requirements outlined in Chapters 5, 6, and 7.
- Examine the steel studs at corners, intersections, terminals, shelf walls, doors, and borrowed-light frames for positive attachment to floor and ceiling runners. All load-bearing and curtain-wall studs must be attached to the runner on each side and at the top and bottom. All load-bearing studs should sit tightly against the runner web.
- Inspect the spliced steel components for proper assembly. Curtain-wall studs and load-bearing framing should not be spliced.
- See that the steel-stud flanges in the field all face the same direction.
- See that preset door frames are independently fastened to the floor slab and that borrowed-light frames are securely attached to the stud and runner rough-framing at all the jamb anchors.
- Make sure that door and borrowed-light frames are spot-grouted, as required.

Suspended Grillages. To inspect the suspended grillage, proceed as follows.

- Measure the spacing of hangers, channels, and studs to see that they are within allowable limits.

- Check the ends of the main runner and furring channels. They should not be let into or supported by abutting walls and should extend to within 6 inches of the wall to support a furring channel.

- Make sure the furring channel clips are alternated and the furring channel splices are properly made.

- See that the mechanical equipment is independently supported and does not depend on the grillage for support.

- Inspect the construction around light fixtures and openings to see that recommended reinforced channel support is provided.

APPENDIX A
METRIC CONVERSION GUIDE

Steel framers frequently need to calculate areas, coverages, volume, etc. Table A-1 shows the conversion factors for these calculations. The standard method of multiplying dimensions is to first convert a fraction of an inch to its decimal equivalent in feet.

DECIMAL EQUIVALENTS

As you can see from Table A-1, these conversions require calculating areas, coverages, volume, and so on. The standard method of multiplying dimensions is to convert the inch portion of the dimension into its decimal equivalent of a foot. Table A-2 provides equivalent decimal amounts for standard fractions of an inch.

TABLE A.1 Metric Conversion Guide for Steel Framer

Property	To convert from	Symbol	to	Symbol	Multiply by	Remarks
Application rate	U.S. gallon per square	gal (U.S.)/100 ft^2	litre per square metre	litre/m^2	0.4075	=0.4075 mm thick
	U.K. gallon per square	gal (U.K.)/100 ft^2	litre per square metre	litre/m^2	0.4893	=0.4893 mm thick
Area	square inch	in.^{2}	square millimeter	mm^2	645.2	1,000,000 mm^2 = 1m^2
	square foot	ft^2	square metre	m^2	0.09290	
	square	100 ft^2	square metre	m^2	9.290	
Breaking strength	pound force per inch width	lbf/in.	kilonewton per metre width	kN/m	0.175	
Coverage	square foot per U.S. gallon	ft^2/gal	square metre per litre	m^2/litre	0.02454	
	square foot per U.K. gallon	ft^2/gal	square metre per litre	m^2/litre	0.02044	
Density, or mass per unit volume	pound per cubic foot	lb/ft^3	kilogram per cubic metre	kg/m^3	16.02	water = 1000 kg/m^3
Energy or work	kilowatt-hour	k Wh	megajoule	MJ	3.600*	J = W • s = N • m
	British thermal unit	Btu	joule	J	1055	
Flow, or volume per unit time	U.S. gallon per minute	gpm	cubic centimetre per second	cm^3/s	63.09	or 0.0631 litre/s
	U.K. gallon per minute	gpm	cubic centimetre per second	cm^3/s	75.77	or 0.0758 litre/s
Force	pound force	lbf	newton	N	4.448	N = kg • m/s^2
	kilogram force	kgf	newton	N	9.807	

Heat flow	thermal conductance, C	Btu/h·ft²·°F	watt per square metre kelvin	W/m²·K	5.678	3 in./ft = 25%
	thermal conductivity, k	Btu·in./h·ft²·°F	watt per metre kelvin	W/m·K	0.1442	1000 µm = 1mm
Incline	inch per foot	in./ft	percent	%	8.333	
Length, width, thickness	mil	0.001 in.	micrometre	µm	25.40*	1000 mm = 1 m
	inch (up to ~48 in.)	in.	millimetre	mm	25.40*	
	foot (~4 ft and above)	ft	metre	m	0.3048*	
Mass (weight)	ounce	oz	gram	g	28.35	1000 g = 1 kg
	pound	lb	kilogram	kg	0.4536	1000 kg = 1 Mg
	short ton	2000 lb	megagram	Mg	0.9072	
Mass per unit area	pound per square foot	lb/ft²	kilogram per square metre	kg/m²	4.882	
	pound per square foot	lb/ft²	gram per square metre	g/m²	4882	
	pound per square	lb/1000 ft²	gram per square metre	g/m²	48.82	
	ounce per square yard	oz/yd²	gram per square metre	g/m²	33.91	
Permeability at 23°C	perm inch	grain·in./ft²·h·in. Hg	nanogram/pascal second metre	ng/Pa·s·m²	1.459	ng = 10¹² kg
Permeance at 23°C	perm	grain/ft²·h·in. Hg	nanogram/pascal second square metre	ng/Pa·s·m²	57.45	1 grain = 64 mg
Power	horsepower	hp	watt	W	746	W = N·m/s J/s
Pressure or stress	pound force per square inch	lbf/in.² or psi	kilopascal	kPa	6.895	Pa = N/m²
	pound force per square foot	lbf/ft² or psf	pascal	Pa	47.88	

TABLE A.1 Metric Conversion Guide for Steel Framer (Continued)

Property	To convert from	Symbol	to	Symbol	Multiply by	Remarks
Temperature	degree Fahrenheit	°F	degree Celsius	°C	$(t_f - 32)/1.8*$	32°F 0°C
	degree Celsius	°C	kelvin	K	$t_c + 273.15*$	273.15K 0°
Thread count (fabric)	threads per inch width	threads/in.	threads per centimetre width	threads/cm	0.394	
Velocity (speed)	foot per minute	ft/min or fpm	metre per second	m/s	0.005080*	
	mile per hour	mile/h or mph	kilometre per hour	km/h	1.609	
Volume	U.S. gallon	gal (U.S.)	cubic metre	m³	0.003785	or 3.785 litres
	U.K. gallon	gal (U.K.)	cubic metre	m³	0.004546	or 4.546 litres
	cubic foot	ft³	cubic metre	m³	0.02832	
	cubic yard	yd³	cubic metre	m³	0.7646	

*Exact conversion factor

TABLE A.2 Decimals of a Foot

Inch	0	1	2	3	4	5	6	7	8	9	10	11
0	0.0000	0.0833	0.1667	0.2500	0.3333	0.4167	0.5000	0.5833	0.6667	0.7500	0.8333	0.9167
1/16	0.0052	0.0885	0.1719	0.2552	0.3385	0.4219	0.5052	0.5885	0.6719	0.7552	0.8385	0.9219
1/8	0.0104	0.0938	0.1771	0.2604	0.3438	0.4271	0.5104	0.5938	0.6771	0.7604	0.8438	0.9271
3/16	0.0156	0.0990	0.1823	0.2656	0.3490	0.4323	0.5156	0.5990	0.6823	0.7656	0.8490	0.9323
1/4	0.0208	0.1042	0.1875	0.2708	0.3542	0.4375	0.5208	0.6042	0.6875	0.7708	0.8542	0.9375
5/16	0.0260	0.1094	0.1927	0.2760	0.3594	0.4427	0.5260	0.6094	0.6927	0.7760	0.8594	0.9427
3/8	0.0313	0.1146	0.1979	0.2813	0.3646	0.4476	0.5313	0.6146	0.6979	0.7813	0.8646	0.9479
7/16	0.0365	0.1198	0.2031	0.2865	0.3698	0.4531	0.5365	0.6198	0.7031	0.7865	0.8698	0.9531
1/2	0.0417	0.1250	0.2083	0.2917	0.3750	0.4583	0.5417	0.6250	0.7083	0.7917	0.8750	0.9583
9/16	0.0469	0.1302	0.2135	0.2969	0.3802	0.4635	0.5469	0.6302	0.7135	0.7969	0.8802	0.9635
5/8	0.0521	0.1354	0.2188	0.3021	0.3854	0.4688	0.5521	0.6354	0.7188	0.8021	0.8854	0.9688
11/16	0.0573	0.1406	0.2240	0.3073	0.3906	0.4740	0.5573	0.6406	0.7240	0.8073	0.8906	0.9740
3/4	0.0625	0.1458	0.2292	0.3125	0.3958	0.4792	0.5625	0.6458	0.7292	0.8125	0.8958	0.9792
13/16	0.0677	0.1510	0.2344	0.3177	0.4010	0.4844	0.5677	0.6510	0.7344	0.8177	0.9010	0.9844
7/8	0.0729	0.1563	0.2396	0.3229	0.4063	0.4896	0.5729	0.6563	0.7396	0.8229	0.9063	0.9896
15/16	0.0781	0.1615	0.2448	0.3281	0.4115	0.4948	0.5781	0.6615	0.7448	0.8281	0.9115	0.9948

EXAMPLE: Multiply 20' – 10¾" × 42' – 6⅜"

20.8958' (the 10¾" became 0.8958')
× 42.5313' (the 6⅜" became 0.5313')
888.7255 (use 889 sq. ft.)

APPENDIX B
MANUFACTURERS' CATALOGS, CHARTS, AND SYMBOLS

A steel-framing manufacturer's catalog can be a confusing mystery, even to construction contractors already building residential steel-framed homes. The reason for this is simple: there is no standardized method for presenting the material. For example, most manufacturers use calculations based on the 1986 American Iron and Steel Institute (AISI) code. The thicknesses used were the minimums specified by AISI with respect to American Society for Testing and Materials (ASTM) standards.

Some manufacturers' catalogs, however, might show higher values for similar products. This usually is the result of calculations that use steel thicknesses that are heavier than those actually used in production. The following figures are based on minimum design thicknesses for a conservative approach.

To help solve the problem of standardization, we are presenting tables and accompanying information from several of the leading manufacturers' catalogs. By indicating what to look for in this literature, we hope to give you a better understanding of how to select the proper materials for your jobs. Remember that these tables and accompanying information are presented to help you understand how to read a manufacturer's catalog and are not intended for use.

WHAT IS IN A MANUFACTURER'S CATALOG?

Refer to the catalog's table of contents to find the desired section. Each section contains structural-properties and maximum-height tables. The structural section also contains joist tables. The following are general guidelines on how to use the information in each section.

Using the Structural-Properties Section

Structural properties for all members are at the beginning of each applicable section. This section contains the physical properties of individual members such as member depth, flange width, the return and weight per linear foot, moments of inertia, section modules, and the radius of gyration along both minor and major axes.

Selecting Wall Materials

If you are selecting a member and have the height of the span, deflection criteria, and wind load, go to the maximum-height section. Select the applicable wind-load column. Then check in ascending order from smallest size to largest until you locate the desired span. At this point, you can choose the desired spacing and gauge for your particular application.

Selecting Joist Loading

Select the span and then follow the column in descending order until you arrive at the desired load.

SELECTING STUDS AND TRACK

See Figure B.1 for the stud designations and profiles used in the table. Table B.1 gives the gross cross sections and Table B.2 gives the net cross sections. Nomenclature is given in the List of Acronyms.

TABLE B.1 Gross Cross Section

Size & type	Gauge	Weight (lbs./ft.)	Area (sq. in.)	$\overline{X}_G$ (in.)	I_X (in.^{4})	S_X (in.^{3})	R_X (in.)	L_Y (in.^{4})	S_Y (in.^{3})	R_Y (in.)
2½" CWS	20	0.578	0.173	0.416	0.176	0.141	1.009	0.040	0.042	0.479
	18	0.762	0.228	0.410	0.230	0.184	1.003	0.051	0.054	0.473
	16	0.944	0.283	0.404	0.281	0.225	0.997	0.062	0.066	0.468
	14	1.183	0.355	0.396	0.347	0.278	0.990	0.075	0.079	0.459
2½" HDC	20	0.691	0.207	0.495	0.205	0.164	0.997	0.057	0.066	0.525
	18	0.910	0.272	0.489	0.267	0.214	0.991	0.074	0.085	0.520
	16	1.127	0.337	0.483	0.327	0.262	0.985	0.089	0.103	0.514
	14	1.389	0.416	0.475	0.397	0.318	0.978	0.107	0.124	0.507
2½" XCSJ	20	0.752	0.225	0.604	0.232	0.186	1.017	0.085	0.085	0.615
	18	0.991	0.296	0.598	0.303	0.243	1.012	0.110	0.110	0.609
	16	1.228	0.367	0.591	0.372	0.297	1.006	0.134	0.133	0.604
	14	1.516	0.453	0.583	0.452	0.362	0.999	0.161	0.161	0.597

TABLE B.2 Net Cross Section

| Area (sq. in.) | $\overline{X}_N$ (in.) | Net cross section | | | | | R_Y (in.) | I_XW (in.^{4}) | RMW (in.-KIPS) | V_{all} KIPS | R_{int} KIPS | R_{ext} KIPS |
		I_X (in.^{4})	S_X (in.^{3})	R_X (in.)	L_Y (in.^{4})	S_Y (in.^{3})						
0.149	0.485	0.175	0.140	1.086	0.035	0.040	0.484	0.175	2.327	0.894	0.262	0.118
0.196	0.479	0.228	0.183	1.080	0.045	0.051	0.478	0.230	3.304	1.178	0.504	0.205
0.242	0.473	0.279	0.224	1.075	0.054	0.062	0.472	0.281	4.285	1.458	0.822	0.392
0.302	0.465	0.345	0.276	1.068	0.065	0.075	0.465	0.347	5.491	1.827	1.374	0.725
0.180	0.569	0.204	0.163	1.065	0.049	0.063	0.525	0.205	3.243	0.973	0.321	0.139
0.236	0.563	0.266	0.212	1.060	0.064	0.081	0.519	0.267	4.225	1.220	0.611	0.267
0.292	0.557	0.325	0.260	1.054	0.077	0.098	0.514	0.327	7.833	2.398	1.333	0.621
0.360	0.549	0.395	0.316	1.048	0.092	0.117	0.507	0.397	9.517	2.952	2.136	1.079
0.198	0.686	0.231	0.185	1.082	0.074	0.080	0.611	0.232	3.579	0.973	0.321	0.139
0.260	0.680	0.301	0.241	1.076	0.095	0.104	0.605	0.303	4.793	1.280	0.611	0.267
0.322	0.674	0.369	0.296	1.071	0.116	0.126	0.600	0.372	8.899	2.398	1.333	0.621
0.397	0.666	0.450	0.360	1.064	0.139	0.151	0.593	0.452	10.833	2.952	2.136	1.079

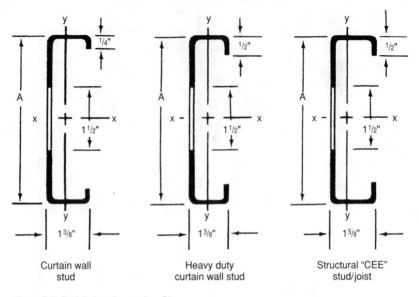

| Curtain wall stud | Heavy duty curtain wall stud | Structural "CEE" stud/joist |

Figure B.1 Stud designations and profiles.

SELECTING WALL COMPONENTS BASED ON WIND LOAD

To select load-bearing wall components that are subject to wind speed, use Table B.3. The table is designed to help with the selection of single-span nonaxial load-bearing wall studs that are subject to uniform lateral wind forces. Select a stud that provides an allowable height equal to or greater than the actual project requirements based on the required spacing, lateral pressure, and deflection criteria. Do not exceed the maximum allowable height.

All calculations in Table B.3 are based on the 1986 edition of AISI's *Specification for the Design of Cold-Formed Steel Structural Members*. The yield strength is 33,000 psi for 20- and 18-gauge members, 40,000 psi for 16-gauge members, and 50,000 psi for 14-gauge members.

The allowable loads and heights are based on stud properties, mechanical bracing on maximum 4-foot centers, and sheathing boards on both flanges properly attached with fasteners on centers of 12 inches or less.

Take into account the following guidelines when selecting load-bearing wall components that are subject to wind load.

- Allowable stresses increase 33 percent for any combination that includes wind loading.
- Fabricate axial-loaded walls with the stud ends seated firmly in the track, which must have full bearing on the structure.
- Load tables are for simple supported spans. Do not use when continuous over supports.
- Minimum 10-inch unpunched steel is required at both ends of the members. When field cutting reduces this minimum, web stiffening might be required.

TABLE B.3 Wind Load

Wind load	Stud size	Deflection	20 ga.	18 ga.	16 ga.	14 ga.
5 psf	8"	L/240	26.0	29.1	31.3	33.7
		L/360	22.7	25.5	27.3	29.5
		L/600	19.2	21.5	23.1	24.9
20 psf	8"	L/240	15.9	18.4	19.7	21.3
		L/360	14.3	16.0	17.2	18.6
		L/600	12.1	13.5	14.5	15.7
30 psf	8"	L/240	13.0	16.0	17.2	18.6
		L/360	12.5	14.0	15.1	16.2
		L/600	10.6	11.8	12.7	13.7
40 psf	8"	L/240	10.0	13.9	15.7	16.9
		L/360	10.0	12.7	13.7	14.7
		L/600	9.6	10.7	11.5	12.4

- Values listed do not account for web crippling. Calculate web crippling per AISI specifications.
- Load-bearing walls might require diagonal bracing and additional wind posts to resist racking.
- Adequate horizontal bridging might be required before the sheathing material is installed. This can be achieved by providing 1½-inch cold-rolled channel (CRC). Attach the channel to each stud by welding or with clip angles and screws. Horizontal strapping and solid bridging with track members also can be used for bridging.
- Attach both stud flanges to the track members at both the top and bottom.
- Design capacities of fasteners and welds shall be per the manufacturer's recommendations.
- All heights are for simple-span conditions only.
- Attach the stud ends to the track components at the top and bottom of the wall assembly.
- Deflections and stresses were calculated without regard to the composite contribution of facing materials.

For components subjected to lateral wind loads of 20 pounds per square foot (psf) and greater, the actual bending and axial stresses were multiplied by 0.75 in accordance with AISI Section A4.4. The 0.75 adjustment was not incorporated into the review of components subjected to a 5- or 10-psf interior lateral load since national codes do not classify them as wind loads. Deflections were calculated on the basis of the actual, unreduced lateral loads shown.

COMBINING AXIAL AND WIND LOADS

To obtain the allowable axial loads in KIPS, or 1000-pound increments, refer to the following general notes and details.

- All calculations are based on the 1986 edition of AISI's *Specification for the Design of Cold-Formed Steel Structural Members.*
- The yield strength is 33,000 psi for 20- and 18-gauge members, 40,000 psi for 16-gauge members, and 50,000 psi for 14-gauge members.

- The allowable loads and heights are based on stud properties, mechanical bracing on maximum 4-foot centers, and sheathing boards on both flanges properly attached with fasteners on centers of 12 inches or less.
- Allowable stresses increase 33 percent for any combination that includes wind loading.
- Fabricate axial-loaded walls with the stud ends seated firmly in the track, which must have full bearing on structure.
- Load tables are for simply supported spans. Do not use when continuous over supports.
- Minimum 10-inch unpunched steel is required at both ends of the members. When field cutting reduces this minimum, web stiffening might be required.
- Values listed do not account for web crippling. Calculate web crippling per AISI specifications.
- All axial-loaded studs must have their ends seated as tightly as possible against the top and bottom tracks.

Provide adequate diagonal bracing, properly designed to prevent wall racking, with crossbraced tension straps. The resulting anchorage and uplift requirements should be adequately compensated for in the design.

DETERMINING ALLOWABLE AXIAL LOADS FOR WALL COMPONENTS

When selecting load-bearing wall components, including braced C-studs, that are subject to uniform lateral and axial loads, follow these guidelines.

- All loads are shown in KIPS.
- Stress increases by 33 percent for both the bending and axial loads where the lateral load exceeds 0.0 psf.
- Members that do not meet an L/240 deflection limit are shown as carrying an 0.0 axial load.
- The following describes the deflection limit nomenclature:
  ~values not followed by (&), (#) or (+) do not exceed L/720
  ~values followed by (+) do not exceed L/600
  ~values followed by (#) do not exceed L/360
  ~values followed by (&) do not exceed L/240
  ~L/240 is exceeded where values have been omitted

DETERMINING ALLOWABLE AXIAL LOADS FOR WEB ELEMENTS

When selecting the axial-loaded internal web elements for trusses, mansard frames, or diagonal kickers and unbraced C-studs that are not subject to lateral loads, refer to Tables B.4 and B.5. Select a stud, in terms of unbraced length, that provides an axial capacity equal to or greater than the applied axial load.

CALCULATING WEB CRIPPLING STRENGTH

To determine the web crippling capacity of a single unreinforced web subject to concentrated loads or reactions, compare the applied concentrated load or reaction to the maxi-

TABLE B.4 Allowable Axial Loads

	20 psf (exterior)				25 psf (exterior)			
	(1⅝" flg)				(1⅝" flg)			
Gauge	20	18	16	14	20	18	16	14
Height (ft)	12" O.C. spacing				12" O.C. spacing			
8	2.05	3.07	5.31	7.11	1.82+	2.82	5.07	6.85
9	1.71#	2.66	4.66	6.28	1.44#	2.36+	4.37	5.97
10	1.36#	2.22#	3.98	5.39	1.06&	1.89#	3.66#	5.05#
11	1.02&	1.79#	3.34#	4.56+	0.71&	1.44&	3.01#	4.20#
12	0.72&	1.39&	2.75#	3.81#		1.03&	2.41&	3.45#
14			1.81&	2.61&				2.25&
16								
Height (ft)	16" O.C. spacing				16" O.C. spacing			
8	1.75+	2.74	4.99	6.76	1.46#	2.42+	4.68	6.42
9	1.35#	2.26#	4.28#	5.87	1.02&	1.90#	3.92#	5.47+
10	0.97&	1.78#	3.56#	4.94+	0.61&	1.38&	3.18#	4.52#
11	0.61&	1.33&	2.90#	4.08#		0.91&	2.50&	3.64#
12		0.92&	2.31&	3.33#			1.91&	2.89&
14								

NOTES:

Values shown assume loads are applied concentrically to the stud.

Studs shall be braced against rotation. Install mechanical bridging spaced at intervals not exceeding 4 feet on center maximum.

Stud ends shall be attached to track components at the top and bottom of the wall assembly. Complete bearing is required.

For components subjected to 5 psf and greater lateral wind loads, the actual bending and axial stresses were increased 33⅓ percent in accordance with AISI Section A4.4, Wind or Earthquake Loads. The 33⅓ increase was not incorporated in the review of components subjected to a 0 psf interior lateral load.

Deflections and stresses were calculated without regard to the composite contribution of facing materials.

mum allowable values, in pounds (Pa). When the applied load exceeds the allowable, reinforcement with a web stiffener is required. For members with multiple webs (i.e., boxed or I-shaped headers), compute the Pa for each web and add the results to obtain the allowable load or reaction for the multiple web. Due to the high degree of restraint against rotation of an I-shaped section, web crippling strengths exceeding those shown below are attainable (Table B.6). Consult AISI Specification Section C3.4, equations C3.4-3, 5, 7, and 9.

DETERMINING LOAD IN FLOOR AND ROOF APPLICATIONS

Refer to Table B.7 when selecting floor and roof joists subject to uniform live and dead loads. The *total-load* values in the following table denote the total safe uniformly distrib-

TABLE B.5   Axial Loads

Section									
0.5F	1.0F	1.5F	2.0F	2.5F	3.0F	3.5F	4.0F	4.5F	5.0F
2.83	2.79	2.71	2.61	2.48	2.31	2.09	1.84	1.56	1.29
3.78	3.71	3.59	3.43	3.23	2.99	2.71	2.40	2.06	1.72
7.10	6.90	6.57	6.12	5.55	4.88	4.13	3.32	2.69	2.25
8.86	8.61	8.20	7.64	6.95	6.16	5.29	4.35	3.58	3.03

NOTES:

Values shown assume loads are applied concentrically to the stud. Applications involving lapped framing conditions eccentrically load the stud resulting in weak (Y-Y) axis bending. Moment (Y-Y) = P × dimension D. Framing members subjected to combined axial and bending shall satisfy the interaction equation of AISI Section C5, Combined Axial and Bending.

Values shown were based on the exclusion of mechanical bridging. L/r(lim) = 200. Values have been omitted where L/r exceeds 200.

TABLE B.6   Web Crippling Strengths

Web height (h)	18 ga. bearing width, n (in.)				16 ga. bearing width, n (in.)				14 ga. bearing width, n (in.)			
	1.5	2.5	3.5	6.0	1.5	2.5	3.5	6.0	1.5	2.5	3.5	6.0
3⅝"	327	382	436	572	721	822	922	1174	1230	1373	1515	1872
(358)	598	674	778	1074	1303	1439	1572	2106	2110	2291	2471	3082
	231	269	308	404	514	586	658	837	884	986	1089	1345
	634	651	669	712	1468	1501	1534	1615	2499	2543	2588	2699
4"	322	375	429	562	711	810	910	1158	1217	1358	1499	1852
(4)	591	665	768	1061	1290	1425	1556	2085	2094	2273	2452	3058
	226	263	301	394	505	575	646	822	872	973	1074	1326
	614	631	648	690	1434	1466	1498	1577	2455	2498	2542	2651
6"	291	339	387	508	658	750	842	1072	1147	1280	1413	1746
(6)	550	619	716	988	1221	1349	1473	1974	2007	2178	2350	2931
	197	229	262	344	455	519	583	742	806	900	993	1227
	509	523	537	573	1251	1279	1306	1376	2217	2257	2296	2394

NOTES:

H/t exceeds 200 where values have been omitted. When h/t exceeds 200, web stiffeners at concentrated load or reaction points are required.

Components with unreinforced flat webs, subjected to combined bending and web crippling, shall be designed to meet the requirements of the following interaction equation: $1.2 (P/Pa) + (M/Ma) \leq 1.5$ where

P = Concentrated load or reaction in the presence of bending moment
Pa = Allowable concentrated load or reaction in the absence of bending moment
M = Applied bending moment at, or immediately adjacent to, the application of the concentrated load or reaction
Ma = Allowable bending moment if bending alone exists

Avoid locating a web punch-out within the shaded areas shown below. Should a punch-out be located in this area, install a web stiffener, minimum 6" wide, centered over the punch-out.

TABLE B.7 Uniformly Distributed Load-Carrying Capacities for Floor and Roof Systems

Total load spacing			Live load spacing		
12"	16"	24"	12"	16"	24"
90	68	45	83	62	42
155	116	77	103	77	52
193	144	96	128	96	64
267	201	134	178	134	89
158	119	79	158	119	79
299	224	149	247	185	123
372	279	186	308	231	154
519	390	260	429	322	215

NOTES:

Applications involving multiple spans, cantilevers, concentrated loads, impact loading, etc. should be investigated separately.

Web crippling should be investigated in accordance with AISI Section C3.4. Web stiffeners are recommended at all support and concentrated load locations.

Joists shall be restrained against rotation at each end. Joists shall be attached to track components or restrained by the installation of continuous solid blocking.

Minimum end bearing shall be 1½ inches.

Deflections and stresses were calculated without regard to the composite contribution of facing materials.

Joist assemblies shall be braced against rotation. Install mechanical bridging spaced at the following intervals:
- Five feet on center maximum for 1⅝ inch flanged components (SW).
- Seven feet on center maximum for all remaining conditions.

uted load-carrying capacities in psf. The *live-load* values denote the live load in psf that can produce a deflection of ⅟₃₆₀ of the span.

Live loads that produce a deflection of ⅟₂₄₀ of the span can be obtained by multiplying the live-load values by 1.5. In no case, however, should the *total-load* capacity of the joists be exceeded. The use of the following table is limited to applications that involve simply supported components installed to a maximum slope of ½ inch per foot.

SELECTING FLOOR AND CEILING JOISTS

Use Tables B.8 and B.9 to determine the joist spans and the maximum allowable spans for uniformly distributed loads.

TABLE B.8. Joist Spans

	6"				8"			
Span	18	16	14	12	18	16	14	12
9'	132.2	235.5	314.2	*	195.2	348.4	*	*
	132.2	195.1	240.6	*	195.2	348.4	*	*
10'	107.1	190.7	254.5	363.5	158.1	282.2	381.3	*
	107.1	142.2	175.4	240.1	158.1	281.8	348.3	*
11'	88.5	157.6	210.3	300.4	130.7	233.2	315.2	*
	86.7	106.8	131.8	180.4	130.7	211.8	261.7	*
12'	74.4	132.5	176.7	252.4	109.8	196.0	264.8	378.7
	66.8	82.3	101.5	138.9	109.8	163.1	201.6	277.2
13'	63.4	112.9	150.6	215.1	93.6	167.0	225.6	322.7
	52.5	64.7	79.8	109.3	93.6	128.3	158.5	218.0
14'	54.6	97.3	129.8	185.4	80.7	144.0	194.6	278.2
	42.0	51.8	63.9	87.5	80.7	102.7	126.9	174.5
15'	47.6	84.8	113.1	161.5	70.3	125.4	169.5	242.4
	34.2	42.1	52.0	71.1	67.6	83.5	103.2	141.9
16'	41.8	74.5	99.4	142.0	61.8	110.2	149.0	213.0
	28.2	34.7	42.8	58.6	55.7	68.8	85.0	116.9
17'	37.1	66.0	88.1	125.8	54.7	97.6	132.0	188.7
	23.5	28.9	35.7	48.9	46.5	57.4	70.9	97.5

TABLE B.9 Maximum Allowable Spans for Uniformly Distributed Loads

		6"			8"			6"		
Span	Gauge	18	16	14	18	16	14	18	16	14
FT	FY	33	50	50	33	50	50	33	50	50
4	TL	589			616			647		
	LL	589			616			647		
6	TL	262	490	602	399	748		287	532	662
	LL	262	490	602	399	748		287	532	662
8	TL	147	276	338	224	421	518	161	299	372
	LL	147	230	282	224	421	518	161	253	311
10	TL	94	176	216	143	269	331	103	191	238
	LL	94	117	144	143	240	295	103	129	159
12	TL	65	122	150	99	187	230	71	133	165
	LL	55	68	83	99	138	171	60	74	92

TABLE B.9   Maximum Allowable Spans for Uniformly Distributed Loads (Continued)

Span	Gauge	6"			8"			6"		
		18	16	14	18	16	14	18	16	14
FT	FY	33	50	50	33	50	50	33	50	50
14	TL	48	90	110	73	137	169	52	97	121
	LL	34	42	52	70	87	107	38	47	58
16	TL	36	68	84	56	105	129	40	74	93
	LL	23	28	35	47	58	72	25	31	38

NOTES:

All loads are computed based on 1986 edition of AISI's *Specification for the Design of Cold-Formed Steel Structural Members.*

Total load values are based on stress limitations, and deflection values are based on deflection limitations of L/240 and L/360.

All values shown are for single span condition only. Compression flange must be braced against lateral buckling.

All joists must be checked for web crippling, in accordance to AISI specification, which may necessitate the use of web stiffening at all reactions, and concentrate loads.

Joist must have a minimum of 10 inch unpunched web at bearing points.

Lateral bracing to consist of cut to length runner track for solid blocking and steel straps on both flanges of joists is necessary to prevent joists from rotating or twisting.

Solid blocking is welded or screw attached between outer joists, over all interior support and adjacent to openings at maximum 8 feet O.C. Strap bracing is screw attached on top and bottom joist flange between solid blocking. Install bridging immediately after joists are erected and before construction loads are applied to prevent flange rotation, and to support flange in compression.

Top flange bridging can be assumed to be provided by the proper attachment of flooring material. Spacing of bridging must be calculated based on actual stress.

Loads shown as TL represent safe uniformity distributed total loads in pounds per linear foot (lb/ft) based on bending stress and shear stress considerations. Loads shown as LL represent live loads in lb/ft based on deflection of L/360 and limited to total loads. All loads are for one span condition, assuming compression flanges to be continuously supported.

The normal hole pattern for structural members used as joists consists of oval holes 1½ inches wide by 4 inches long at 12 inches from one end and 24 inches O.C. thereafter. The data shown in the joist tables assumes that there are no holes in the vicinity of high bending moments and/or high shear.

Provide bridging or lateral support as needed. If the compression flange is continuously supported, the tension flange must be laterally supported at least at 8 ft c/c, using straps or other suitable bridging. Where diaphragm-rated elements such as plywood or metal deck are to be attached to the compression flange, loads may not be imposed on the joists until completion of the installation of the diaphragm elements.

INCORPORATING FIRE-RATED ASSEMBLIES

Table B.10 depicts various fire-rated assemblies that incorporate lightweight steel-framing components. Rather than listing all the specifications (e.g., attachment requirements, assembly constraints, etc.) we suggest that you research the applicable standard through the agency that conducted the test.

TABLE B.10 Fire-Rated Assemblies

Test reference	Fire rating	Type of assembly	Agency	Components
FM24676.4 FC224	2 hr.	Floor/ceiling	FM 1975	• 2½" concrete • 9⁄16" 28 ga. deck and mesh • 7¼ × 18 ga. joists, 24" O.C. • 2 layers ⅝" G.W.B. ceiling
FM29135 FC245	1 hr.	Floor/ceiling	FM 1977	• 2" concrete, (Note B) • 1 5⁄16", 24 ga. deck • 6 × 18 ga. joists, 24" O.C. • 1 layer ½" G.W.B. ceiling
L524	1 hr.	Floor/ceiling	UL 1988	• Min. 7¼ × 18 ga., steel stud, 24" O.C. • Use any of the floor systems indicated in the UL test.
P511	1 hr.	Roof/ceiling	UL 1988	• Min. 7¼ × 18 ga., steel joist, C shape, 2-inch flange minimum, 24" O.C.
P512	1 hr.	Roof/ceiling	UL 1988	• Min. 7¼ × 18 ga., steel joist, C shape, 24" O.C.
U418	¾ hr.	Bearing wall	UL 1988	
U418	1 hr.	Bearing wall	UL 1988	• Two layers ½" thick, G.W.B., one side • 3½ or 5½ × 18 ga. steel stud, 24" O.C.
U418	2 hr.	Bearing wall	UL 1988	• Three layers ½" thick G.W.B., one side • 3½ or 5½ × 18 ga. steel stud, 24" O.C.
U425	¾, 1 hr.	Bearing wall interior	UL 1988	
U425	1½ hr.	Bearing wall interior	UL 1988	• Two layers ½" thick, G.W.B., each side • 3½ × 20 ga. steel stud, 24" O.C.
U425	2 hr.	Bearing wall interior	UL 1988	• Three layers ½" thick G.W.B. • 3½ × 20 ga. steel stud, 24" O.C.
U425	¾, 1, 1½ hr.	Bearing wall exterior	UL 1988	
U425	2 hr.	Bearing wall exterior	UL 1988	• Three layers ½" thick G.W.B., interior side • 3½ × 20 ga. steel stud, 24" O.C.
U426	3 hr.	Bearing wall	UL 1988	• Four layers ½" thick G.W.B., each side • 3½ × 20 ga. steel stud, 24" O.C.
U434	1 hr.	Bearing wall	UL 1988	• ⅞" thick portland cement plaster • 3½ × 20 ga. steel stud, 24" O.C. • One layer ⅝" thick G.W.B. interior

NOTES:
A UL denotes Underwriters Laboratories, Inc., and FM denotes Factory Mutual Research Corporation
B Lightweight concrete measured from top flute of deck
C G.W.B. (gypsum wallboard)

APPENDIX C
CONNECTOR AND FASTENER INFORMATION

The connector and fastener information given in Tables C-1 through C-9 can be used in conjunction with that given in Chapter 3.

Based on data provided in Chapter 3, the following are suggestions for the proper use of screws in the three basic construction phases of a steel-framed home. Keep in mind that these suggestions are meant to provide a comparative analysis only and must *not* be used for design applications. Consult the manufacturer's data, catalogs, values, and specifications.

TABLE C.1 Suggested Capacity for Screw Connections (Pounds)

Gauge	Fy (KSI)	No. ¼-14 D = 0.188" T = 0.205"		No. 12-14 D = 0.160" T = 0.177"		No. 10-16 D = 0.138" T = 0.153"		No. 8-18 D = 0.120" T = 0.125"		No. 6 D = 0.100" T = 0.106"	
		Shear or bracing	Tension	Shear or bracing	Tension	Shear or bracing	Tension	Shear or bracing	Tension	Shear or Bracing	Tension
12	50	886	533	682	491	495	476	N/A	N/A	N/A	N/A
14	50	774	367	624	326	433	311	N/A	N/A	N/A	N/A
16	50	645	241	571	232	395	229	358	215	N/A	N/A
18	33	301	101	276	101	263	98	248	94	188	83
20	33	N/A	N/A	N/A	N/A	141	69	140	68	133	53

NOTES:

N/A—not applicable—two thicknesses of this gauge metal should not be connected with this size screw.

Screw spacing and edge distance shall not be less than ½ inch.

Screw capacities are based on average test results divided by a safety factor of 3.0.

When connecting materials of two different gauges, use the value of the thinner material.

TABLE C-2 Suggested Capacities for Welded Connections (Pounds/Inch)

Gauge	Fy (KSI)	Fu (KSI)	Fillet welds		Flare-groove welds	
			Longitudinal loading	Transverse loading	Longitudinal loading	Transverse loading
12	50	65	2384	2644	1983	2203
14	50	65	1594	1854	1890	1644
16	50	65	1212	1472	1104	1226
18	33	45	631	811	609	676
20	33	45	467	623	467	519

NOTES:

Values are based on AISI Specification Section E2.

When connecting materials of two different thicknesses, use the values for thinner material.

Although values for 20-gauge material is given in this table, welding of 20-gauge materials is only recommended under controlled conditions.

Good welds can be obtained with ³⁄₃₂ or ⅛ inch diameter, type E-60 or E-70 welding rods. Shop welding is best accomplished with 0.030- to 0.035-inch-type E70 wire.

TABLE C-3 Suggested Design Loads for Powder-Driven Fasteners in Concrete (Pounds).

Shank diameter	Minimum penetration	Type of loading	Concrete compression strength (psi)		
			2000	3000	4000
0.145"	1⅛"	Tension	70	135	145
		shear	130	200	215
0.177"	1½"	Tension	120	245	315
		shear	200	325	385
0.205"	1¼"	Tension	175	225	275
		shear	310	355	400

TABLE C-4 Suggested Bearing Capacity when Used to Connect Gauge Thickness Steel (Pounds)

Shank diameter	Steel gauge thickness	
	18 ga.	20 ga.
0.145"	—	—
	—	225
0.177"	—	—
	359	275
0.205"	—	—
	—	319

NOTES:

Capacities shown are for cured-stone aggregate concrete. Minimum base thickness of concrete shall be 1.3 × depth of embedment of fastener.

Minimum fastener spacing: 4 inches; minimum fastener edge distance: 3 inches.

Bearing capacities based on the 1986 AISI Specification Section E3.3. Fu = 45 KSI per ASTM A446, Grade A (Fy [min] = 33 KSI).

TABLE C-5 Suggested Design Loads in Tension or Shear for Powder-Driven Fasteners in Structural Steel (Pounds)

Cold-rolled steel gauge	0.145" diameter Hot-rolled steel thickness			0.177" shank diameter Hot-rolled steel thickness			0.205" shank diameter Hot-rolled steel thickness		
	¼"	⅜"	½"	¼"	⅜"	½"	¼"	⅜"	½"
12	210	210	210	335	395	395	485	525	660
14	210	210	210	335	395	395	485	525	581
16	210	210	210	335	395	395	465	522	522
18	210	210	210	335	359	359	416	416	416
20	210	210	210	275	275	275	320	320	320

NOTES:

Tests were conducted with the fastener point driven completely through the back side of the hot-rolled steel member. This is necessary to obtain proper gripping force.

Bearing capacities based on the 1986 AISI Specification Section E3.3. $F_u = 45$ KSI per ASTM A446, Grade A (F_y [min] = 33 KSI).

TABLE C-6 Recommended Fasteners in Floor Framing

Location	Fastener	Frequency or quantity
Joist to girder	½" #10 panhead	1 @ joist to girder
Joist to connection clip	½" #10 panhead	3 to 4 @ each clip
Bridging to joist	½" #10 panhead, hexhead	1 @ each joist
Joist to 2× wooden end stiffener	1¼" #10 teks pancake head	
End stiffener to joist	¾" #10 panhead	3 to 4 @ each stiffener to joist
End stiffener to wooden rim joist	8d common wire nail	2 @ each end stiffener to rim joist
Steel rim joist to end stiffener	¾" #10 pan head	3 @ each joist
Steel rim track to end stiffener	¾" #10 panhead	3 @ each joist
Joist hanger to joist	⅞" #10 panhead, hexhead	3 @ each joist
Joist to overlapping joist	¾" #10 panhead	3 @ support
Wooden sole plate (wall) to rim joist & track	2½" #12 flathead	1 @ 24" O.C. & max. 12" from each end of track
Plate track (bottom) to joist & track	1¹⁵⁄₁₆" #12 panhead with pilot point	1 @ 24" O.C. & max. 12" from each end of track
Subfloor to joist	buglehead	1 @ 16 or 24" O.C.
Sheathing to joist	buglehead	1 @ 16 or 24" O.C.
Floor joist to steel beam	hexhead	1 @ 24" O.C.
Rim track to steel beam	hexhead	1 @ 24" O.C.
Solid blocking	panhead	1 @ 16 or 24" O.C.

TABLE C-7   Recommended Fasteners in Load-Bearing Wall Framing

Location	Fastener	Frequency or quantity
Stud to plate track (bottom)	¾" #8 low-profile panhead	1 @ each flange
Stud to plate track (top)	¾" #8 low-profile panhead	1 @ each flange
Diagonal bracing to stud	½" #8 low-profile panhead	1 @ each stud
Lateral bracing to stud	¾" #8 panhead	1 @ each stud per strap or 3 @ each connection clip with cold-rolled channel
Gusset to stud	¾" #10 low-profile panhead	Quantity and spacing as per loading
Stud to stud (nested)	¾" #8 panhead	1 @ 24" O.C. through flange
Stud to stud (back to back)	¾" #8 panhead	1 @ 24" O.C. through web
Stud to stud (@ wall intersection)	¾" #10 panhead	1 @ 24" O.C. or 1 @ each blocking
Lintel to stud	¾" #10 panhead	Requirement varies with different loading
Drywall to stud	1" #10 type-S buglehead	1 @ 16" or 24" O.C.
Sheathing to stud	1" #10 type-S buglehead	1 @ 16" O.C.
Furring to lath	½" #8 type-W buglehead	1 @ 16" O.C.
Rigid insulation to stud	1½" #10 type-S buglehead	1 @ 16" O.C.
Ladder-back to stud	¾" #8 panhead	1 @ 24" O.C.
Windowbrace to stud	¾" #10 panhead	1 @ each stud per strap
Window sill to stud	¾" #8 panhead	1 @ 24" O.C.
Wood cabinet to stud	2½" #10 ovalhead	1 @ 16" O.C.

TABLE C-8   Recommended Fasteners in Nonload-Bearing Wall Framing

Location	Fastener	Frequency or quantity
Stud to plate track (bottom)	½" #8 screws low-profile panhead	1 @ each flange
Stud to plate track (top)	½" #8 screws low-profile panhead	1 @ each flange
Lateral bracing to stud	½" #8 screws panhead	2 @ flange
Stud to stud (nested)	½" #10 screws panhead	1 @ 24" O.C.
Stud to stud (back to back)	½" #10 screws panhead	1 @ 24" O.C.
Stud to stud (@ wall intersection)	½" #10 screws panhead	1 @ 24" O.C. or 1 @ each blocking
Drywall to stud	1" #10 type-S buglehead	1 @ 16" or 24" O.C.
Sheathing to stud	1" #10 type-S buglehead	1 @ 24" O.C.
Furring to lath	½" #8 type-W buglehead	1 @ 24" O.C.
Rigid insulation to stud	1½" #10 type-S buglehead	1 @ 16" O.C.

TABLE C-9   Recommended Fasteners in Roof Framing

Location	Fastener	Frequency or quantity
Ceiling joist to wooden top plate	1" #12 panhead	1 @ each joist
Ceiling to joist to top plate track	¾" #10 panhead	1 @ each joist
Connection clip to wooden top plate	1" #12 pancakehead	4 @ each clip to top plate
Connection clip to top plate track	¾" #10 panhead	4 @ each clip to plate track
Connection clip to ceiling joist	¾" #10 panhead	Min. 3 @ each clip to ceiling joist and as per loading
Connection clip to rafter	¾" #10 panhead	Min. 3 @ each clip to rafter and as per loading
Ceiling joist to parallel rafter	¾" #10 panhead	No. varies as per loading
Ceiling joist to truss web	¾" #10 panhead	Min. 2 @ flange and as per loading
Ceiling joist, overlapped at support	¾" #10 panhead	Min. 2 @ web
Connection clip to ridge board	¾" #10 panhead	4–6 @ each clip to ridge
Rafters overlapped at ridge	¾" #10 panhead	Min. 6 @ overlapped web section and as per loading
Built-up beam (ridge board)	¾" #10 panhead	1 @ each flange @ 12" O.C.
Stiffback bracing to joist	¾" #10 panhead	Min. 2 @ each joist
Subfascia track to rafter	¾" #10 low-profile panhead	1 @ each connection clip and max top plate
Wooden fascia to subfascia track	1⅝" #12 trimhead	2 @ 24" O.C. and @ maximum of 12" from each end of board or corner
Rafter to rafter	¾" #10 panhead, hexhead, buglehead	Min. 2 @ 24" O.C.
Collar tie to rafter	¾" #10 panhead, hexhead	2 @ per each collar tie
Bridging to rafter	¾" #10 panhead, hexhead	2 @ min.
Rafter to ceiling joist	¾" #10 panhead, hexhead	1 @ each joist
Gusset to rafter	¾" #10 panhead	Min. 3 @ gusset
Kingpost to rafter	1" #10 panhead	1 Min. each rafter
Truss web to rafter	1¼" #10 panhead	No. varies per loading
Bracing to rafter	¾" #10 panhead, hexhead	2 @ each rafter
Rafts to web stiffener	¾" #10 panhead	1 @ each rafter
Trim molding to rafter	¼" #8 ovalhead	1 @ 24" O.C.
Drywall to rafter	1" #10 type-S buglehead	1 @ 16" O.C.
Rigid insulation to ceiling joist	1" #10 type-S buglehead	1 @ 24" O.C.
Drywall to ceiling joist	1" #10 type-S buglehead	1 @ 16" O.C.

NOTES:
   Low-profile panhead screws can be used in lieu of panhead screws where least projection of the fastener is desired.
   S-7 point can be used as a substitute for S-12 when attaching 0.07-inch members together.
   Where wood is fastened to steel channels, common wire nails and screw shank nails can be used in place of screws.

APPENDIX D
MATERIAL TAKE-OFF FORM

As discussed in Chapter 4, the square footage of a job is good primarily for *guesstimate*. The material take-off form described in this appendix, on the other hand, is the first in a series of detailed steps that you should follow in order to arrive at a more realistic estimated cost of construction. It is time-consuming because it looks at (and causes you to analyze and make decisions about) each component that will be used to build a house. Spend this time and effort, however, and your chances for a successful and profitable project improve greatly.

The following instructions offer a detailed guide for taking off the material requirements from the job plans or blueprints. The instructions follow the same format as the material take-off form and are divided into the same three sections:

- Foundation and concrete slab
- Exterior shell
- Interior

Further, each material type is discussed in six categories:

- Description
- Use
- Estimating procedure
- Example
- Accuracy rate
- Scrap consideration

ESTIMATING FOUNDATIONS AND CONCRETE SLABS

Cushion Sand

Description. Depending on the area of the country and local construction practices, cushion sand material can be river sand, washed sand, bank gravel, or washed gravel. A minimum 4-inch thickness is recommended. Place the sand after the grade beams are dug.

Use. Cushion sand helps obtain a uniformly flat surface to ensure that the concrete slab is of a consistent thickness. The sand absorbs, to some minor degree, the uplift pressure of soil expansion and help dissipate the water in the concrete during the curing process.

Estimating Procedure. Cushion sand is not placed at grade beams, therefore figure the square footage of the slab area minus the square footage of the grade beams. (Grade beams are 1 foot 0 inches wide. To obtain the square footage of the beam, multiply the beam length by 1 foot.) Then multiply the remaining square footage by the 4-inch thickness of the sand and divide by 27, which is the unit by which the sand is sold (27 cubic feet equals one cubic yard).

Example

         2162   sq. ft. of slab area
    −   481   sq. ft. of grade beam
         1681   sq. ft. of cushion sand
    × 0.3333   ft. thick (decimal equivalent of 4 inches thick)   .
       560.28   cubic feet
    ÷   27
        20.75   cubic yards (round to 21 cubic yards)

Accuracy Rate. Because of the possible unevenness of the dirt below the sand, this amount can be from 10 to 15 percent.

Scrap Consideration. If too much sand is ordered, use the excess for the flat work, such as walks and driveways, or landscaping beds.

Re-Bar

Description. Re-bar is reinforcing steel. The number designation is the diameter of the rod in eighths of an inch, e.g., #5 re-bar is ⅝ inch in diameter, #10 re-bar is 1¼ inches in diameter. The bar has ridges, or is *deformed* to help it bond to the concrete.

Use. Re-bar is used in grade beams to add strength to the beam. Re-bar is used in the slab to keep cracks from enlarging. When concrete cures it shrinks, thus causing minor cracking. Because re-bar helps keep cracks from enlarging, it is sometimes called *shrink-steel.*

1. #5s ×20'0" Long = Used at Grade Beam

Estimating Procedure. Multiply linear feet of grade beam by 4, which is the number of #5s in each grade beam. Then divide the total by 19. The re-bar is 20 feet 0 inches long, but each piece should lap 1 foot 0 inches.

Example

        481   lin. ft. of grade beam
    ×   4   (4 pieces per beam)
       1924   lin. ft.
    ÷   19
     101.26   pieces (use 102 pieces)

Accuracy Rate. The accuracy rate of this estimating method is 97 to 103 percent.
Scrap Consideration. Construction practices and poor use of drops can reduce accuracy to (–)5 to 10 percent.

2. #3s × 20'0" Long = Used at Slab

Estimating Procedure. Method #1. Divide square footage of slab area by 14 to get the number of 20-foot-0-inch-long pieces needed.

Method #2. Using the foundation drawing, lay out each piece to scale, showing 12-inch laps and splices, and indicate where drops are used.

Accuracy Rate. Method #1 is 95 to 105 percent accurate. Method #2 is about 98 to 102 percent accurate.

Scrap Consideration. Poor use of drops can reduce accuracy by (–)10 percent.

Optional Use of Welded Wire Mesh. Use 6 × 6/10.10 welded wire mesh in lieu of re-bar at the slab.

Description. The designation 6 × 6/10.10 indicates that the mesh is a 6-x-6-inch grid, using 10-gauge wire in both directions.

Use. As an alternate to the #3s, the mesh is the more common slab-reinforcement material in some regions of the country. Local construction practices dictate its use.

Estimating Procedures. Divide square footage of slab area by 5 feet 6 inches. The mesh is 6 feet 0 inches wide, but sidelaps 6 inches.

Example

2162   sq. ft. of slab
÷ 5.5   (decimal equivalent of 5'6")
 393   lin. ft. of mesh

Accuracy Rate. The accuracy rate of this estimating method is about 102 percent.
Scrap Consideration. This method does not use all of the drops.

3. #3s Stirrups

Description. #3, or ⅜-inch-diameter, stirrups is re-bar purchased prebent as shown on the material take-off form. Stirrups do not come in standard sizes, therefore, you need to specify the size to your vendor. In this example, the specification is for #3 × 6 inch × 17 inch.

Use. #3 stirrups are used in the grade beams to hold the main re-bars in place and to transfer loads between the top and bottom re-bars. They are located every 18 inches along each grade beam.

Estimating Procedure. Divide the total length of all grade beams by 1 foot 6 inches, to determine the number of pieces.

Example

  481   lin. ft. of grade beam
÷   1.5   (decimal equivalent of 1'6")
320.66   pieces (use 321 pieces)

Accuracy Rate. Depending on placement and construction practices, the accuracy rate is from 98 to 102 percent.
Scrap Consideration. This consideration is not applicable.

Vapor Barrier

Description. 6 mil-, or 0.066-inch-, thick polyethylene film, variously called poly, visqueen, or paper. The vapor barrier material is purchased in 10-foot-wide rolls of various lengths.

Use. Placed above the sand and below the concrete slab, the vapor barrier is used to retard the flow of water vapor up from the subsoils through the concrete slab. Its omission, in wet weather, shows as dampness on the surface of the slab. This dampness can cause vinyl tile adhesives to soften, carpets to mildew, and generally adds excessive amounts of water vapor to a room. The disadvantage to a vapor barrier, however, is that it prevents the sand cushion from removing some of the water from the concrete during the curing process. This causes all of the excess concrete water to pass through the top surface, which causes a weak, approximately ⅛-inch-thick film to form on the surface. While this weak surface layer is structurally harmless, it is subject to chipping and spalding from dropped objects. Therefore, one rule of thumb is to use a vapor barrier wherever there is floor covering and to omit it where there is not (i.e., such as in the garage).

Estimating Procedure. Calculate the square footage of the slab area and then subtract the garage and porch area, if desired.

Example. An example is not necessary.

Accuracy Rate. The accuracy rate should be 100 percent.

Scrap Consideration. This consideration is not applicable.

Anchor Bolts

Description. Generally, anchor bolts are ⅝ inch in diameter × 12 inches long, with a 2-inch hook. That is, one leg is 12 inches long and the other leg is 2 inches long for a total leg length of 14 inches. The projection above the slab is 2 inches in a 20-inch embedment. Therefore the anchor bolt should have 2 inches of thread.

Use. Accurately set, with a template, anchor bolts are used to hold the columns to the foundation.

Estimating Procedure. Count them several times as shown on the anchor-bolt sheet.

Example. An example is not necessary.

Accuracy Rate. With careful counting, the accuracy rate should be 100 percent.

Scrap Consideration. Add 2 or 3 to your count to safeguard against thread damage, etc.

Concrete

Description. All structural concrete shall have a minimum 28-day compressive strength of 3000 psi, 3- to 5-inch slump, and a maximum aggregate size of ¾-inch hardrock.

NOTE: The use of 3000 psi concrete in some regions of the country, such as California, requires continuous onsite inspection by the local building authority while the concrete is being poured. The required inspection can be costly and can be eliminated by having a local architect or engineer design the foundation system using 2000 psi concrete.

Use. Concrete is used for the foundation and slab.

Estimating Procedure. Divide the square footage of the slab area by 81. Multiply the linear feet of the grade beams and multiply by 0.074. Add the two totals together.

Example

```
  2162   sq. ft. of slab
÷   81
  26.69  cu. yds.
   481   lin. ft. of grade beam
× 0.074
  35.59  cu. yds.
  26.69  cu. yds.
+ 35.59  cu. yds.
  62.28  cu. yds.
```

Accuracy Rate. Because of actual job site conditions, such as variations in grade beams, spills, etc., you should expect an accuracy rate of no better than 95 to 105 percent.

Scrap Consideration. Try not to have any. The last truck is going to dump the balance of his/her load on your job site which, of course, you pay for because you ordered it. One common practice is to place the last few yards on *will-call*. The second from the last truck radios the batch plant and orders exactly what is needed to finish the pour. Coordinate this procedure with the concrete company when you place your initial order.

CALCULATING MATERIALS FOR EXTERIOR SHELL

Roof

The two numbers you need to calculate the various components of the roof are the perimeter length of the roof, which is the total linear feet of all of the roof edges, and the square footage of the roof surface.

Calculating the perimeter length
Method 1: Gable ends
a. Gable endwalls or rakes
　　From the floor plan, find the dimension from the steel line to the peak. To this number, add 2 feet 4½ inches, which is an average of all slopes for the horizontal projection of the eave overhang. Multiply this number by the appropriate factor in Table D-1. This answer is the dimension from the edge of the roof to the peak. Repeat this process for all of the sloped-roof edges shown on the elevations.

　　Example
　　20'0"　(steel line to peak on floor plan)
　　+ 2'4½"　(eave overhang)
　　22.3750　(decimal equivalent of 22' 4½")
　　× 1.054　(slope factor for 4/12 pitch)
　　23'7"　(edge of roof to peak)

b. Eaves, including all horizontal roof edges
　　From the floor plan, find the length of that section of the roof and add 2 feet 1¼ inches for each gable overhang shown on the elevations. NOTE: 2 feet 1¼ inches is measured from the centerline of the last frame to the roof edge.

c. Total all dimensions to arrive at perimeter length

TABLE D.1
Calculating Slope

Slope	Factor
3/12	1.031
4/12	1.054
5/12	1.083
6/12	1.118
7/12	1.158
8/12	1.202
9/12	1.250
10/12	1.302

Method 2: Square footage of roof surface
Measure all roof edges shown on the elevations, and total the amounts.
CAUTION: Not all roof edges, such as porch insets, show on the elevations.

Calculating the square footage of the roof surface
Start with the square footage of the slab, which was calculated in the foundation section, and then subtract the square footage of the garage apron and any porches or patios not covered by the roof. Multiply the resultant area by the slope factor in Table D-2 and add twice the perimeter length to arrive at the square footage of the roof surface. This formula works even if the roof has hips and valleys.
Example

	2042	sq. ft. (slab area)
–	32	sq. ft. (garage apron)
–	148	sq. ft. (patio not under roof)
	1862	sq. ft.
	× 1.047	(slope factor for combination 3/12 and 4/12)
	1949.5	
+	710	(twice the enterprise roof perimeter of 355 lin. ft.)
	2659.5	sq. ft. of roof surface

Accuracy Rate. The accuracy rate for this formula is 95 to 102 percent.

1. 30# felt × 36" wide × 72' long roll
Description. This is asphalt impregnated felt. The 30# designation indicates the shipping weight per square (100 sq. ft.).
Use. Asphalt impregnated felt is used as an underlayment to shingles, and as an aid in waterproofing the roof.
Estimating Procedure. It is assumed that each roll has one 6-inch sidelap and two 6-inch endlaps, giving a net coverage of 2 feet 6 inches × 71 feet, or 177.5 square feet per roll. Therefore, divide the square footage of the roof by 177.5 and round up to the next full roll.
Example

2656	sq. ft. of roof area
÷ 177.5	(coverage per roll)
14.96	rolls (use 15 rolls)

Accuracy Rate. For a simple roofline without hips and valleys, accuracy should be 100 percent.
Scrap Considerations. A roof with hips, valleys, and complex architectural treatment can cause a 5- to 15-percent scrap factor.

TABLE D.2 Roof Surface
(Square Feet)

Slope	Slope factor
4/12	1.047
Combination of 3/12 & 4/12	1.047
6/12	1.149

2. Shingles

Description. There are a multitude of roofing products on the market, including asphalt, clay tile, slate, metal, etc. At this point a decision must be made as to the type of shingle to be used and manufacturer's literature consulted for details of installation.

Use. These instructions are not the proper forum for a full discussion on roofing systems. Suffice it to say that because of the relatively horizontal nature of a roof, it is subject, more than any other segment of a house, to environmental abuse. In spite of this, the roof is a house's first line of defense against rain and wind damage to the home's interior. It behooves the professional builder to become well versed on roofing products and their proper use.

Estimating Procedure. Roofing is estimated and sold in squares. A square of roofing is the amount required to cover 100 square feet. Divide the square footage of the roof surface by 100.

Example

$$
\begin{array}{l}
2655 \quad \text{sq. ft. (surface area of roof)} \\
\underline{\div 100} \\
26.55 \quad \text{(use 27 square)}
\end{array}
$$

Accuracy Rate. The accuracy rate should be 98 to 102 percent.

Scrap Consideration. Add about 10 percent to the estimated amount for waste. For hips and valleys, or roofs with complex architectural treatment, the waste can be as much as 15 percent.

3. Metal flashings
a. Drip edge

Description. The drip edge is a galvanized sheetmetal angle installed around the perimeter of the roof to prevent rainwater from wicking back up under the shingles.

Use. Install a drip edge along all roof edges.

Estimating Procedures. The amount of drip edge required is equal to the length of the roof perimeter calculated earlier.

Example. An example is not required.

Accuracy Rate. The accuracy rate should be 100 percent.

Scrap Consideration. This consideration is not applicable.

b. Valley flashing with 20-inch metal roll flash

Description. Valley flashing is a light-gauge, galvanized steel or aluminum (in some rare roof systems, copper is specified) roll of material, which is cut to length and formed in the field.

Use. Install flashing at the junction of the two roof planes in the valley. Place beneath the shingles to provide waterproofing.

Estimating Procedure. You have already calculated or scaled the distance from the edge of the roof to the peak. Multiply that dimension by 1.4 to obtain the length of the valley or hip.

Example

$$
\begin{array}{l}
23.5833' \quad (23'7'' \text{ edge of roof to peak}) \\
\underline{\times \ 1.414} \\
33'4\tfrac{5}{16}'' \quad \text{(length of valley)}
\end{array}
$$

Accuracy Rate. The accuracy rate should be 99 to 101 percent.

Scrap Consideration. Allow about 12 inches for cut-ins at the top and bottom of the flashing run.

c. 6-inch metal roll flashing

Description. Roll flashing is a light-gauge, galvanized steel or aluminum (in some rare roof systems, copper is specified) roll of material, which is cut to length and formed in the field.

Use. At the junction of a roof and high wall, install the flashing behind the siding and above the shingles, to provide waterproofing.

Estimating Procedures. Using the elevations on the plan set, scale the length of all roof/high-wall junctions, and total the quantities.

Example. An example is not required.

Accuracy Rate. The accuracy rate is 95 to 105 percent.

Scrap Consideration. Allow about 12 inches for cut-ins at the top and bottom of the flashing run.

d. Metal shingles (step flashing)

Description. Metal shingles are light-gauge galvanized steel or aluminum (in some rare roof systems, copper is specified). Shingles are available in two configurations: 5-×-7-inch flat for field bending and 4-×-4-inch angle × 8 or 10 inches long.

Use. Use to flash a brick chimney into the roof. They are installed in the mortar joint by the brick mason, then field bent by the roofer and fastened to the roof between the shingles.

Estimating Procedure. Determine the perimeter dimension of chimney, then divide by the length of the step flashing chosen. Allow about a 1½-inch lap at each piece.

Example

12'0"	(perimeter of chimney)
÷ 6.5	(decimal equivalent of 8" step flashing less 1½" lap)
22.15	pieces (use 23 pieces)

Accuracy Rate. The accuracy rate should be 100 percent.

Scrap Consideration. Add a few to protect against damage in field.

4. Fasteners

Description. Fasteners are galvanized roofing nails with a large round or square sheet-metal washer. The washer prevents the felt from tearing over the nail.

Use. Install 30# felt at the rate of 50 nails per square.

Estimating Procedure. Multiply the number of squares of shingles by 50, then divide by 100, which is the quantity per pound, to arrive at the number of pounds required.

Example. An example is not required.

Accuracy Rate. The accuracy rate should be 100 percent.

Scrap Consideration. Add one pound for waste.

b. Shingle nails

Description. The type and quantity of fasteners is determined by the type of shingles used. Refer to the shingle manufacturer's installation instructions.

Cornice

There are many definitions for the term cornice. The one adopted for the material take-off is as follows:

All soffit and fascia material, and all exterior wooden trims, including those above and below the windows.

1. Soffit

Description. The soffit is the underside of the eave and gable overhangs. Typically, the soffit material is masonite or plywood with a smooth or rough-sawn finish and a thickness of ¼ inch, ⅜ inch or ⁷⁄₁₆ inch.

Use. In addition to being an architectural finish to the underside of the overhangs, soffits seal air leaks that would occur if it were omitted.

Estimating Procedure. Soffit material comes in 4-x-8-foot sheets. Each sheet cuts lengthwise to provide 16 linear feet of soffit material. Divide the roof perimeter length by 16 to obtain the number of sheets required.

Example

```
 355  lin. ft. (roof perimeter)
÷  16
21.875  (use 23 sheets)
```

Accuracy Rate. This estimating procedure generally provides one or two sheets too many.

Scrap Consideration. Short drops and unexpected cuts probably will use up the extra sheets.

2. Fascia

Description. The finishing edge to the overhangs is the fascia. Typically, the material is a 1 × 10. Note that the actual area to be covered is 8⅛ inches.

Use. Use fascia as an architectural finish to the edge of the overhang, as well as to seal air and water leaks.

Estimating Procedure. The total linear feet of fascia required is equal to the perimeter length of the roof.

Example. An example is not required.

Accuracy Rate. The accuracy rate should be 100 percent.

Scrap Consideration. Because of the probability of short drops and unexpected cuts, add 5 to 10 percent.

3. Shingle mold

Description. Material can be 1-x-2-inch, 1-x-3-inch, or 1-x-4-inch finger-jointed redwood or other soft wood.

Use. Shingle mold is used as a trim piece to the fascia board and as a support to the drip edge.

Estimating. The total linear feet of the required shingle mold is equal to the calculated perimeter length of the roof.

Example. An example is not required.

Accuracy Rate. The accuracy rate should be 100 percent.

Scrap Consideration. Because of the probability of short, unusable drops and unexpected cuts, add 5 to 10 percent.

4. Batten strips at soffit joints

Description. Material is ¼-x-¾-inch screen molding.

Use. Batten strips are used as trim at the butt joints between pieces of the plywood used at the soffit.

Estimating Procedure. There is no exact procedure. Generally, allow 2 feet per each end joint of soffit material. There are two end joints per piece of soffit material.

Example

```
23  sheets of soffit material
×4  (two 2'0" pieces per sheet)
92  lin. ft. of batten strip
```

Accuracy Rate. Because of the probability of short fill-in pieces of soffit, this method might be short by 15 percent.

Scrap Consideration. Add about 15 percent.

5. Soffit to siding trim
a. Continuous 2 × 4 blocking at brick

Description. The siding trim is 2 × 4 blocking installed continuously at the soffit to receive brick-frieze trim board. Used only when siding is brick veneer.

Use. Its purpose is to provide a nailing surface for brick-frieze trim board.

Estimating Procedure. The amount of 2 × 4 blocking is equal to the total linear feet of brick veneer around the house. Using the plans, calculate this total length.

Example. An example is not required.

Accuracy Rate. The accuracy rate should be 100 percent.

Scrap Consideration. Because of the probability of short, unusable drops and unexpected cuts, add 5 to 10 percent.

b. Trim/brick frieze

Description. Material can be 1-×-3-inch or 1-×-4-inch finger-jointed redwood or other soft wood.

Use. Serves as brick frieze when used in conjunction with brick veneer or as soffit-to-siding trim when used in conjunction with any other siding. NOTE: In many cases when brick is used, there are places over the windows and gable areas where siding also is used. While the 2 × 4 blocking is used only at brick areas, the trim is continuous around the perimeter of the house.

Estimating Procedure. The amount of trim is equal to the total perimeter dimension of the house, including any and all areas where the soffit joins an exterior wall. Using the plans and elevation, calculate and/or measure the total length.

Example. An example is not necessary.

Accuracy Rate. The accuracy rate should be 100 percent.

Scrap Consideration. Because of the probability of short, unusable drops and unexpected cuts, add 5 to 10 percent.

6. Window and door trims

Description. All of the various trims can be used in conjunction with all types of exterior siding except brick. A variety of materials can be used; the materials frequently are dictated by the type of siding used (i.e., vinyl trim pieces with vinyl siding, etc.).

Use. Window and door trims are used as an architectural finish to cover raw edges of the siding, to seal air and water leaks, and as a smooth surface to receive caulking.

Estimating Procedure. Using the elevations from the plan set, calculate and/or measure the lengths of the various trims. Also refer to the garage door-jamb detail shown on the detail sheet. Include enough length for mitered corners.

Example. An example is not needed.

Accuracy Rate. Depending on the accuracy of the scale of the drawings, the accuracy rate should be about 100 percent.

Scrap Consideration. Because of the probability of short, unusable drops, add about 10 percent.

7. Porch soffit
a. Soffit

Description. The soffit is the underside of the porch roof. Select the same material used for the overhang soffits.

Use. The soffit is used as an architectural finish to the underside of the porch roof, and to seal air leaks that would occur if it were omitted.

Estimating Procedure. Soffit material comes in 4-×-8-foot sheets. Using the floor plan page of the plan set, lay out the soffit using a 4-×-8-foot grid, in such a manner as to hold splices to a minimum. Count the number of sheets.

Example. An example is not needed.

Accuracy Rate. The accuracy rate should be 100 percent.

Scrap Consideration. Unless a sheet is mitre-cut, you do not need extra material.

b. Soffit batten strips

Description. Material is ¼-×-¾-inch screen molding.

Use. Use it as a trim at butt joints between pieces of soffit material.

Estimating Procedure. From the layout done for the porch soffit, calculate the linear feet of exposed butt joints.

Example. An example is not needed.

Accuracy Rate. The accuracy rate should be 100 percent.

Scrap Consideration. Because of the probability of short, unusable drops, add 5 to 10 percent.

c. Soffit to siding trim

Description. Material can be 1-×-3-inch or 1-×-4-inch finger-jointed redwood or other soft wood.

Use. Use as brick frieze when used in conjunction with brick veneer or as soffit-to-siding trim when used in conjunction with any other siding.

Estimating Procedure. Amount required is equal to the length of the porch soffit perimeter.

Example. An example is not needed.

Accuracy Rate. The accuracy rate should be 100 percent.

Scrap Consideration. Because of the probability of short, unusable drops, add 5 to 10 percent.

8. Wooden nailer

Description. A wooden nailer generally is 2 × 10 lumber.

Use. Use a wooden nailer at endwalls of two story homes. It also is used to provide a nailing surface outboard of the second-floor beam. Refer to the endwall section on the detail sheets of the plan set.

Estimating Procedure. The total length of material required is twice the house width.

Example

40'0" (width of house)

× 2

80 lin. ft. required

Accuracy Rate. The accuracy rate should be 100 percent.

Scrap Consideration. This consideration is not applicable.

9. Garage door framing

Description. Door-framing material is 2 × 4 lumber.

Use. As a continuous block at jambs and header of garage door, garage door framing is used only in conjunction with brick veneer.

Estimating Procedure. Calculate the total length of jambs and header of garage door.

Example. An example is not needed.

Accuracy Rate. The accuracy rate should be 100 percent.
Scrap Consideration. This consideration is not applicable.

C. Siding

The number you will be using to calculate the exterior veneer (siding, brick, etc.) is the total square footage of all exterior walls less the square footage of doors and windows. The calculations are divided into two parts:

* All of the area below the soffit line
* The gable areas

The area below soffit line is defined as the intersection of the roof overhang soffit and the exterior wall. This line is extended as an imaginary line across the gabled endwalls of the house. There are five variables that affect the height of the exterior veneer (siding):

* Whether the rafters are 8- or 10-inch-deep channels
* Whether the finish ceiling is flat or cathedral (sloped)
* Whether or not a brick ledge is used
* Whether the house is single story or two story
* The roof slope

Use Table D-3 in concert with the two footnotes, to arrive at the height of the exterior veneer (siding). Because not all plan sets indicate whether the rafters are 8 or 10 inches, the table reflects an average of the two rafter depths. This shortcut introduces an error in the overall accuracy of less than 1 percent.

Example. Using the rear wall and assuming a brick veneer, use this process with all below-soffit-line areas to arrive at a total area.

```
    7.6435   (4/12 roof slope with a sloped ceiling)
 +  0.1250   (brick ledge)
 +  9.7083   (two story)
   17.4687   (17'5¾" total wall height)
 × 33.4167   (33'5" wall length)
  583.7463   (use 584 sq. ft. of wall area)
```

TABLE D.3 Exterior Sheathing Heights

Roof slope	Flat ceiling		Sloped ceiling	
	Height	Decimal equivalent	Height	Decimal equivalent
3/12	8'9¼"	8.7708	7'10¼"	7.8542
4/12	8'6⅝"	8.5521	7'7⅝"	7.6354
6/12	8'1¾"	8.1458	7'2¾"	7.2292
8/12	7'9⅛"	7.7604	6'10⅛"	6.8438
9/12	7'6⅞"	7.5729	6'7⅞"	6.6563
10/12	7'4⅝"	7.3854	6'5⅝"	6.4688

If brick is used add 1½" (0.1250)
If wall is two story, add 9'8½" (9.7083)

From the plans for the gable areas, determine the horizontal distance from the sidewall to the peak and add 2 feet 0 inches for the overhang. Do not confuse this measurement with the 2-foot-4½-inch measurement used earlier. We now are measuring to a point 7 inches below the roof plane. Multiply this dimension by the appropriate slope factor from Table D-4 to determine the rise. Multiply the horizontal dimension by the rise for the area of a gabled wall. Divide the answer by 2 if the house has a shed roof.

Example

18'0"	(sidewall to peak)
+ 2'0"	(overhang)
20'0"	
× 0.3333	(rise factor for 4/12 roof)
6.666'	(total rise to peak)
× 20'0"	
133.32	(use 133 sq. ft. for gabled roof)
÷ 2	(for shed roof)
66.66 sq.ft.	(use 67 sq. ft. for shed roof)

Continue this process with all gable areas to arrive at a total area.

NOTE: The sum of the areas below the soffit line and the gable areas is the total surface area of the exterior walls. From this total, deduct the square footage of all doors and windows. The answer is the total amount of veneer required.

C. Siding
1. Synthetic plaster

Description. Synthetic plaster is a generic term used to describe any of a variety of insulated or noninsulated plaster or stucco systems. It is generally purchased installed as a subcontract because of the high degree of technical expertise required for its installation.

Use. Synthetic plaster is used as an architectural veneer and as a weathering surface for protection of the house.

Estimating Procedure. Synthetic plaster is purchased by the square foot, installed. The amount is equal to the square feet of veneer calculated earlier, less openings for windows and doors.

Example. An example is not needed.

TABLE D.4
Calculating Rise

Slope	Rise factor
3/12	0.25
4/12	0.3333
5/12	0.4167
6/12	0.5
7/12	0.5833
8/12	0.6667
9/12	0.75
10/12	0.8333

Accuracy Rate. Because it is bid and purchased as a subcontract, use your figures only as a check against the subcontractor's bid.

Scrap Consideration. This consideration is not applicable.

2. Brick

Description. One of the more commonly used brick for residential applications is the standard modular, which has a nominal dimension of 2⅔ × 8 inches and lays-up at the rate of 6.75 bricks per sq. ft. There are many sizes, colors, and face textures available. Brick sample cards are available from brick suppliers to aid in your selection.

Use. While brick has structural value it rarely is used for load-bearing purposes on a house. The brick on a home is laid-up as a veneer wall. The brick-ties securely affix the brick to the sheathing, but in a fashion that does not transfer any load.

Estimating Procedure. Because of the variety of brick sizes, these instructions cannot presume to provide the number of brick required. Given the square footage of wall, however, your brick supplier can provide quantities needed based on the chosen brick size.

Example. An example is not needed.

Accuracy Rate. Depending on the accuracy of the square footage calculations, the rate should be between 97 and 103 percent.

Scrap Consideration. Add about 5 percent for waste, breakage, etc.

3. Siding

Description. A variety of sidings are available, the most common being wood, aluminum, and vinyl. They are available in a variety of widths. Again, the most common being 5, 6, and 8 inch. Typically, aluminum and vinyl siding is installed as a system; that is, they require special trim and support pieces designed specifically for the siding. If the house is to receive these kinds of siding, refer to the manufacturer's literature for details.

Use. Siding is used as an architectural finish to the house and as a weathering surface to prevent the infiltration of the environment, such as wind and rain, into the house. It also contributes to the sheathing's ability to resist the racking forces of winds.

Estimating Procedure. If aluminum or vinyl siding is selected, refer to the manufacturer's literature for estimating procedures and details.

Use the following information to figure linear feet.

- 4-inch-wide siding: 3 lin. ft. = 1 sq. ft.
- 5-inch-wide siding: 2.4 lin. ft. = 1 sq. ft.
- 6-inch-wide siding: 2 lin. ft. = 1 sq. ft.
- 8-inch-wide siding: 1.5 lin. ft. = 1 sq. ft.

Example. An example is not necessary.

Accuracy Rate. Depending on the accuracy of the sq. ft. calculations, the rate should be between 97 and 103 percent. Also refer to scrap considerations below.

Scrap Consideration. Because of the probability of short, unusable drops and bevel cuts at gable areas, add 3 to 5 percent.

D. Windows

Description. Windows shown on the standard plan sets are aluminum-framed with double glazing, and the size is specified. Materials, sizes, quantities, and styles are all subject to change depending on you or your customer's preference.

Use. Windows add light and ventilation.

Estimating Procedure. Count them.

Example. An example is not needed.

Accuracy Rate. The accuracy rate should be 100 percent.
Scrap Consideration. This consideration is not applicable.

E. Miscellaneous exterior materials
1. Lintels
Description. Lintels are made from hot-rolled steel angle. Recommended sizes are based on the weight of a standard modular brick with a thickness of 4 inches and a dead load weight of 40 pounds per square feet of surface area. Heavier masonry veneers require larger lintels.

Use. Lintels support brick veneer over windows and doors. The lintels are installed by the masons as they lay the brick. The lintel requires 4 inches of bearing at each end; that is, the lintel must be 8 inches longer than the width of the opening. They are used only in conjunction with brick or other masonry units and are not required with siding. Unless there is a cutting torch on the job site, you probably want to purchase them cut-to-length.

Estimating Procedure. One lintel is required for each door or window that has masonry above it. Tally door and window openings, by width and add 8 inches to each.

Example. An example is not needed.

Accuracy Rate. The accuracy rate should be 100 percent.

Scrap Consideration. Some vendors charge for the drops from a 20-foot stock length.

2. Gable vents
Description. Vents with light-gauge, fixed-blades that are nonadjustable come in the sizes and shapes shown on the drawings and can be wooden or metal.

Use. Installed in the gable endwall, vents allow air exchange into the attic space to prevent heat buildup in warm weather and water-vapor condensation in winter.

Estimating Procedure. Refer to the elevations on the drawings and count the vents.

Example. An example is not needed.

Accuracy Rate. The accuracy rate should be 100 percent.

Scrap Consideration. This consideration is not applicable.

3. Brick flashing
Description. The most commonly used material is 20-mil visqueen (PVC) that is 12 to 18 inches wide.

Use. Moisture buildup in the cavity behind a brick veneer can occur because: water vapor pressure can force water vapor to migrate from the interior of the house through the wall and condensate on the cool back surface of the brick. Rainwater, especially a driven rain, can penetrate through minute cracks between brick and mortar. Install the brick flashing at the base of the wall to direct water drainage out through the weep holes.

Estimating Procedure. The required linear feet of brick flashing is equal to the total linear feet of brick veneer.

Example. An example is not needed.

Accuracy Rate. The accuracy rate should be 100 percent.

Scrap Consideration. Any lapped splice works against the flashing's efficiency, therefore allow enough to avoid splices.

4. Skylights
Description. Any of a variety of windows designed can be located on the roof. Generally, fixed, but operable types are used. Most typically, skylights are constructed of an aluminum frame with a flat or domed plastic lite. The better models have roof flashing integral to the frame. They are available in a variety of sizes.

Use. Skylights are used to admit natural light from above.

Estimating Procedure. Count them.
Example. An example is not needed.
Accuracy Rate. The accuracy rate should be 100 percent.
Scrap Consideration. This consideration is not applicable.

5. Crickets
  Description. Also called a saddle or water diverter, a cricket is installed on the roof at the upslope side of a chimney to divert rainwater around the chimney. Not available as a purchased item, you need to either field-fabricate them from light-gauge metal or special order them from a sheetmetal shop.
  Use. A cricket is used to divert rainwater around the upslope side of a chimney.
  Estimating Procedure. Allow one per chimney.
  Example. An example is not needed.
  Accuracy Rate. The accuracy rate should be 100 percent.
  Scrap Consideration. This consideration is not applicable.

ESTIMATING INTERIOR MATERIAL REQUIREMENTS

Drywall

Residential drywall installations typically can be purchased three ways:

- As a complete materials and labor subcontract
- The general contractor purchases the gypsum wallboard and then purchases, as a sub-contract, the secondary materials, such as joint compound, tape, screws, etc., and installation labor.
- The general contractor purchases all of the material and either subcontracts the labor or hires the labor by the hour.

1. Gypsum wallboard
  Description. Gypsum wallboard commonly is called sheetrock or drywall. The name sheetrock is actually U.S. Gypsum's brand name for gypsum wallboard. The most common thickness for residential construction is ½ inch. Some building codes, however, require a 1-hour fire-separator wall between the living portion of the house and the garage. In these cases, the most common wallboard is ⅝ inch thick, fire-rated. Verify the construction practices with the local building inspector or fire marshall.
  Use. In addition to providing a smooth surface to receive paint or wall covering to finish a wall, the wallboard acts as a structural element to stabilize the metal or wooden studs. While other sizes are available, the most common size is a 4-×-8-foot sheet.
  Estimating Procedure. Divide the square footage of the floor area of the house by 10. Then divide the square footage of the surface area of the vaulted portion of the walls by 0.32.
  NOTE: This procedure only works with 8-foot ceilings. Calculate any wall surface above 8 feet as you did the vaulted portion of the walls.

Example

2700 sq. ft. (sq. ft. of house including garage)
÷ 10
270 pieces
1428 sq. ft. (surface area of
÷ 32 vaulted portion of walls)
44.63 (use 45 pieces)
45 PCs
270 pieces
+ 45 pieces
315 pieces

Accuracy Rate. The accuracy rate is between 90 and 110 percent, depending on installation practices and the use of scrap.

Scrap Consideration. Preplanning, especially with vaulted walls, can reduce scrap somewhat.

2. Ready-mix joint compound

Description. Generally, ready-mix joint compound is a vinyl-based, thin-putty-consistency compound sold in waterproof boxes or plastic packs.

Use. Use ready-mix joint compound to bed wallboard tape and spot nailheads.

Estimating Procedure. Figure 5.75 gallons of joint compound per 1000 square feet of surface area.

Example

315 PCs (4-x-8-foot drywall)
× 32 (sq. ft. in one sheet)
10080 sq. ft.
÷ 1000
10.08
× 5.75
57.96 (use 12 5-gallon pails)

Accuracy Rate. Depending on application expertise, the accuracy rate should be about 100 percent.

Scrap Consideration. This consideration is not applicable.

3. Joint tape

Description. Joint tape is a perforated paper tape that is feathered on the edges and purchased in rolls.

Use. Joint tape is used in conjunction with joint compound to seal wallboard joints.

Estimating Procedure. Figure 370 linear feet of tape per 1000 square feet of wall surface.

Example

10,080 sq. ft.
÷ 1000
10.8
× 370
3729.6 lin. ft. of tape
÷ 250 (typical roll length)
14.92 (use 15 rolls)

Accuracy Rate. The accuracy rate should be 95 to 105 percent.

Scrap Consideration. This consideration is not applicable.

4. Drywall screws

Description.   Use type-S buglehead × 1¼ inch long. The type-S point is designed to pierce up to 20-gauge steel. The buglehead is designed to self-countersink into the wallboard without tearing the face paper.

Use.   Use to attach wallboard to metal studs. Apply with a screwgun that operates at 0 to 4000 rpm.

Estimating Procedure.   Figure 1065 screws per 1000 sq. ft. of wall surface. 258 screws equal one pound.

Example
```
10,080   sq. ft.
÷ 1000
  10.8
× 1065
10,735   screws
÷  258
  41.61   (use 42 lbs.)
```

Accuracy Rate.   When installed as outlined, the accuracy rate should be 100 percent.

Scrap Consideration.   A lot of them are going to get dropped. Add 15 percent for waste.

5. Metal corner bead

Description.   A metal corner bead is a 1¼-×-1¼-inch light-gauge angle that is 8 feet long. Both legs are perforated to receive drywall screws and joint compound.

Use.   Used at all exposed outside corners, the metal corner bead protects the edge as well as provide clean straight lines.

Estimating Procedure.   Calculate the linear feet of exposed outside corners and divide by 8.

Example.   An example is not needed.

Accuracy Rate.   The accuracy rate should be 100 percent.

Scrap Consideration.   Avoid splices when possible. Add 5 to 10 percent because of short, unusable drops.

Miscellaneous Framing

1. Furnace platform

Description.   Furnaces that are located in a closet might require a platform, in a size recommended by the manufacturer, that acts as a return-air plenum. If a furnace located in the attic is not suspended from the rafters, you might need to build a platform on which to place the unit. Additionally, if the furnace is not located near the attic access, you might need to build a service walkway to it. Either platform can be constructed with 2 × 4s and ¾-inch plywood.

Use.   See the description for possible uses.

Estimating Procedure.   There is no procedure. Based on the location, determine the amount of 2 × 4s and plywood required.

Example.   An example is not needed.

Accuracy Rate.   The accuracy rate cannot be determined.

Scrap Consideration.   This consideration is not applicable.

2. Attic access

Description.   Attic access generally is a hatchway into the attic space, with or without pull-down stairs. Reinforce, frame-out, and trim the ceiling area around the access hole.

Framing can be 2 × 4s, trim can be base or casing trim, and door can be plywood or scrap wallboard. If pull-down stairs are used, the door face will be part of the pull-down stair unit.

Use. See the description for various uses.

Estimating Procedure. There is no procedure. Select materials to be used and determine amount.

Example. An example is not needed.

Accuracy Rate. An accuracy rate cannot be determined.

Scrap Consideration. This consideration is not applicable.

3. Prefabricated metal fireplace

Description. Prefabricated metal fireplaces are available in a wide range of styles for burning both wood and gas. Select the model early in the construction schedule so that the manufacturer's installation instructions can be incorporated into the chase construction. It would be wise to also check with the building inspector as regards clearance to combustible materials and other safety features required by the local code.

Use. A prefabricated metal fireplace is an architectural treatment that enhances the salability of a house. Models that incorporate an air-circulation system can be used to lower the house's heating bill.

Estimating Procedure. There is no procedure. Consider the additional materials needed to build the chase, chimney, etc.

Example. An example is not needed.

Accuracy Rate. An accuracy rate cannot be determined.

Scrap Consideration. This consideration is not applicable.

C. Trims

1. Base wall molding

Description. Base wall molding is a molded wooden trim, generally about 3 inches high, available in a variety of patterns. It most commonly is the same pattern, but larger in size, as the door trim. If it is to be painted, finger-jointed material is acceptable, however, staining requires a better grade of material.

Use. Install as a trim piece at the junction of a wall and floor. It is used to cover the raw, unfinished bottom edge of the wallboard. Optionally, it can be used in the garage on more expensive homes.

Estimating Procedure. Unfortunately, there is no formula for calculating the amount required. You can either field-measure after the interior walls are stood or, using the plans, scale each wall, remembering to omit the wall length behind tubs, vanities, and base cabinets. If, when scaling the drawings, you do one room at a time, the process goes quickly and is not too confusing. Do not forget the closets.

Example. An example is not needed.

Accuracy Rate. Depending on your ability to scale the drawings, the accuracy rate should be about 100 percent.

Scrap Consideration. Installing base wall molding generates a lot of scrap. Add about 10 percent. Purchasing the longest length available reduces scrap somewhat.

2. Window stool

Description. The window stool is the interior base of a window and is part of the window trim. The interior edge has a molded pattern. If interior trims are to be stained, you might want to consider a hardwood.

Use. See the description for usage.

Estimating Procedure. From the drawings, determine each window width and add 1½ inches for end revel. Because window stools cannot be spliced, you need to know the available stock lengths and then determine how many pieces can be cut from each stock length.

Example.   An example is not needed.

Accuracy Rate.   The accuracy rate should be 100 percent.

Scrap Consideration.   This is determined when you figure how many pieces can be cut from the stock lengths.

3. Window stool trim

Description.   Window stool trim is molded wooden trim, generally the same pattern as base wall molding and door trim or casing, but of a width equal to the door trim.

Use.   It is used as a decorative trim to the window steel and to cover the raw edge of the drywall below the window.

Estimating Procedure.   See procedure for window stool.

Estimate.   An example is not needed.

Accuracy Rate.   See rate for window stool.

Scrap Consideration.   See consideration for window stool.

4. Utility shelving

Description.   For use in closets, pantries, or any shelving not generally exposed. The material can be dimensional lumber or particleboard.

Use.   Generally used for any shelving not exposed.

Estimating Procedure.   From the drawings, determine how many of each length is required. Because shelving cannot be spliced, you need to know the available stock lengths. Then determine how many pieces can be cut from each stock length.

Example.   An example is not required.

Accuracy Rate.   The accuracy rate should be 100 percent.

Scrap Consideration.   This is determined when you figure how many pieces can be cut from the stock lengths.

5. Decorative/adjustable shelving

Description.   Exposed shelving, such as book shelves, back bars, etc. Material can be veneered wood, solid hardwood, or, in the case of back bars, it can be glass.

Use.   See the description for possible uses.

Estimating Procedure.   Refer to utility shelving.

Example.   Refer to utility shelving.

Accuracy Rate.   Refer to utility shelving.

Scrap Consideration.   Refer to utility shelving.

6. 1 ×3 utility shelving support

Description.   1-×-3-inch finger-jointed redwood or similar soft wood usually is used.

Use.   The supports usually are located at each end and across the back of each shelf.

Estimating Procedure.   The back pieces are equal to the total length of the shelving and the end pieces are equal to two times the width of each shelf.

NOTE: This material can be spliced, but only at a wall stud.

Example.   An example is not needed.

Accuracy Rate.   The accuracy rate should be 100 percent.

Scrap Consideration.   Because splices can occur only at studs, add 10 percent.

7. 1¼-inch-diameter closet rods

Description.   Closet rods are smooth, 1¼-inch-diameter soft wooden dowels.

Use.   Closet rods are located in closets as shown on the drawings.

Estimating Procedure.   Closet rods cannot be spliced. Scale the drawings for the lengths required, then determine how many pieces can be cut from each stock length.

Example. An example is not needed.

Accuracy Rate. The accuracy rate should be 100 percent.

Scrap Consideration. This is determined when you figure how many pieces can be cut from each stock length.

Doors

1. Entry doors

Description. A vast range of styles are available in both wooden and metal doors. The selection of the main entry door might be of major concern to the owner. Some codes require special considerations for entry doors, such as door thickness and locking devices.

Use. Doors are used for access and security.

Estimating Procedure. Refer to the door schedule in the plan set for the size and description of the required doors and to the floor plan for quantities.

Example. An example is not needed.

Accuracy Rate. The accuracy rate should be 100 percent.

Scrap Consideration. This consideration is not applicable.

2. Interior passage doors

Description. Interior passage doors are typically wood and either hollow- or solid-core. Depending on application and room space, they can be swing, bifold, or sliding. Consider solid-core doors for additional privacy at master bedrooms and bathrooms. Prehung doors reduce the in-place cost per unit.

Use. Interior passage doors provide access and privacy.

Estimating Procedure. Refer to the door schedule in the plan set for sizes and descriptions and to the floor plan for quantities.

Example. An example is not needed.

Accuracy Rate. The accuracy rate should be 100 percent.

Scrap Consideration. This consideration is not applicable.

Floor Coverings

Typically, floor coverings are purchased installed as a subcontract item. As such, you need to know the amount involved (square yards, for carpet; square feet for tile), but quantities of mastic, grout, etc., are of little concern because they are included in the unit price. Because these items are available in such a wide price range and are generally selected after the contract is signed, it is common to include their cost as an allowance.

1. Carpet

Description. Carpet is a heavy fabric floor covering, made from a variety of fibers in a variety of pile thickness and densities. Some carpets are installed over a pad and others are glued directly to the floor.

Use. As floor coverings, carpets ideally are located away from areas that have a high incidence of spills, such as kitchens, and dirt, such as entryways.

Estimating Procedure. All carpet, to one degree or another, is directional. Therefore, install the entire house in the same direction. Unfortunately, this does not allow you to minimize the amount of scrap. Additionally, if the carpet has a pattern, you will generate a scrap when the pattern is aligned at a splice. There are two methods for estimating carpet:

1. Using the floor plan

Lay out each area and take note of large pieces of scrap that can be used elsewhere. Then determine how many linear feet of a 12-foot-wide roll are needed and then multiply the linear feet of the roll used by the carpet width and divide by 9 to obtain square yards. NOTE: Some carpets are available in 15-foot-wide rolls.

Example

$$
\begin{array}{r}
160 \quad \text{lin. ft. of 12'-wide carpet} \\
\times \quad 12 \\
\hline
1920 \quad \text{sq. ft.} \\
\div \quad 9 \\
\hline
213.3333 \quad \text{(use 214 sq. yds.)}
\end{array}
$$

2. Quick estimate

Calculate the square footage of the areas to be carpeted, but do not deduct for walls. Divide by 9 to arrive at square yards.

Example.   An example is not needed.

Accuracy Rate.   The accuracy rate for method 1 is 97 to 103 percent. The rate for method 2 is 92 to 108 percent.

Scrap Consideration.   The amount of scrap is determined using method 1.

2. Vinyl tile

Description.   Formerly called vinyl-asbestos tile (VAT); now called vinyl composition tile(VCT), vinyl tile is a common finished flooring material. The most common size is 12 × 12 inches and the most common residential thicknesses are $\frac{1}{16}$ and $\frac{3}{32}$ inch.

Use.   Vinyl tile is a common floor covering.

CAUTION: Floor preparation for VCT is critical. Plaster spills, etc., must be removed and chipped concrete must be patched. If installed over wood, patch nailheads and plywood joints. Over time, VCT flows and something as small as a nailhead depression can show in 3 to 6 months.

Estimating Procedure.   VCT is sold in boxes of 45 pieces each. Each box equals 45 square feet. Using the finished room schedule and the floor plan, scale or calculate the total square footage required and divide by 45.

Example

$$
\begin{array}{r}
81 \quad \text{sq. ft.} \\
\div 45 \\
\hline
1.8 \quad \text{(use 2 boxes)}
\end{array}
$$

Accuracy Rate.   The accuracy rate should be about 100 percent.

Scrap Consideration.   Small areas, such as the example, can generate as much as 10 percent scrap. Areas that cover several hundred square feet can generate as little as 5 percent scrap.

3. Sheet vinyl

Description.   Sheet vinyl is a floor covering composed principally of polyvinyl chloride and should not be confused with linoleum, which is made with cork or wood with a linseed oil binder. Sheet vinyl is available in 6-, 9-, and 12-foot-wide rolls, depending on the manufacturer.

Use.   See the description for usage.

Estimating Procedure.   Like carpet, sheet vinyl is directional and should be estimated as you did the carpet.

Example.   An example is not needed.

Accuracy Rate.   The accuracy rate should be 97 to 103 percent.

Scrap Consideration. The layout determined by the estimate identifies the amount of scrap. However, the pattern repeat needs to be considered when estimating the layout.

4. Ceramic Tile.

Description. Ceramic tile is a hard, durable floor tile made from clay with a ceramic, or metal oxide, glaze and then fired at high temperatures. It is available in a large variety of sizes, patterns, and colors.

CAUTION: Some tile, most notably quarry tile, requires a recess in the concrete slab. Refer to the manufacturer's installation instructions for details and the depth of the recess.

Use. See the description for usage.

Estimating Procedure. Using the finished room schedule and the floor plan, scale or calculate the total square footage required.

Accuracy Rate. The accuracy rate should be 100 percent.

Scrap Consideration. Small areas can generate as much as 10 percent scrap; larger areas about 5 percent.

Wall Coverings

Wall coverings, like floor coverings, are typically purchased installed as a subcontract. Like floor coverings, they are available in such a wide price range it is common to include their cost as an allowance.

1. Wallpaper

Description. Wallpaper is available in such a vast array of colors, patterns, materials, roll widths, and pattern repeats that a serious estimate should not be attempted until the wallpaper has been selected. In order to develop an allowance, however, calculate the square footage.

Use. See the description for usage.

Estimating Procedure. From the floor plans and elevations of the kitchen and bath, scale the wall area to be covered. Do not deduct for windows and doors.

Example. An example is not necessary.

Accuracy Rate. The accuracy rate should be about 100 percent.

Scrap Consideration. Scrap amounts are controlled almost entirely by the type of wallpaper selected.

2. Ceramic tile

Description. Ceramic tiles are 4¼-×-1¼-inch tiles installed over waterproof drywall or cement-enriched plaster. They can be installed up to 70 inches above the finished floor or full height.

Use. They are used in showers and tubs.

Estimating Procedure. For tubs up to 70 inches above the finished floor (AFF), allow 44 square feet per tub. For the tub's full height, allow 65 square feet per tub. For showers, scale or calculate perimeter and multiply by 5.8333 for 70 inches AFF or multiply by 8.000 for full height.

Example. An example is not needed.

Accuracy Rate. The accuracy rate should be about 100 percent.

Scrap Consideration. This consideration is not applicable.

a. Ceramic soap dish

Description. Molded ceramic soap dish

Use. One per shower or tub

Estimating Procedure. Count them.
Example. An example is not needed.
Accuracy Rate. The accuracy rate should be about 100 percent.
Scrap Consideration. This consideration is not applicable.

b. 24-inch ceramic towel bar
Description. Only the end supports of the towel bar are ceramic. The bar is generally plastic.
Estimating Procedure. Count one per tub.
Example. An example is not needed.
Accuracy Rate. The accuracy rate should be about 100 percent.
Scrap Consideration. This consideration is not applicable.

Millwork

1. Wooden stairs
Description. Stairs can be functional or architectural. The treads can be carpeted or hardwood. The risers can be open or closed. The job carpenters generally can build the more functional stairs, but a millwork shop might be required to build and install architectural stairs.
Use. See description for usages.
Estimating Procedure. There generally are one set per house.
Example. An example is not needed.
Accuracy Rate. The accuracy rate should be 100 percent.
Scrap Consideration. This consideration is not applicable.

2. Cabinets
Description. Upper and base kitchen cabinets, vanity bases, and other cabinets can be purchased prefabricated by the unit or custom-made by the linear foot. When they are custom-made, they are generally installed by the manufacturer rather than the job carpenters or framers.
Use. See the description for usage.
Estimating Procedure. From the floor plans and elevations of the kitchen and bath, scale or calculate the total length.
NOTE: To determine the total length, measure across the face of the cabinet.
Example. An example is not needed.
Accuracy Rate. The accuracy rate should be 100 percent.
Scrap Consideration. This consideration is not applicable.

Tops

Description. Plastic laminate is a generic term for commonly recognized trade names, such as Formica or Wilsonart countertops. Plastic laminate is manufactured from multiple layers of resin-impregnated paper, which is then fused together under heat and pressure. It is laminated to substrata such as plywood or particleboard to form countertops. Alternate materials might be ceramic tile on kitchen countertops and synthetic marble on vanities.
Use. Kitchen countertops, bath vanity tops, and miscellaneous bar tops are likely to be covered with plastic laminate or another appropriate surfacing material.
Estimating Procedure. Using the floor plans and kitchen and bath elevations, scale or calculate the total countertop length.

NOTE: Measure countertops to the outside corner.
Example. An example is not needed.
Accuracy Rate. The accuracy rate should be 100 percent.
Scrap Consideration. This consideration is not applicable.

Glass Shower Door

Description. This item includes both sliding glass tub enclosures and shower fronts. Use either tempered or wire glass.
Use. See the description for usage.
Estimating Procedure. Tub enclosures shown on the plans generally are 5 feet wide.
Example. An example is not needed.
Accuracy Rate. The accuracy rate should be 100 percent.
Scrap Consideration. This consideration is not applicable.

Hardware

Description. Typically, a residential construction contract includes hardware as an allowance because of the vast range of products and price levels. Hinges and drawer guides are not listed because they are furnished as part of the prehung door or cabinet package.
Use. Included are all of the devices used to open doors and drawers, miscellaneous brackets, catches, stops, and secondary bathroom fixtures.
Estimating Procedure. Carefully review the drawings and count each hardware type.
Example. An example is not needed.
Accuracy Rate. The accuracy rate should be 100 percent.
Scrap Consideration. This consideration is not applicable.

Plumbing Fixtures

Description. Generally, these items are part of the plumbing subcontract. Selecting and counting them now aids in developing the plumbing subcontract. Before they can be selected, you need to determine if the hot water heater is to be gas or electric.
Use. See the description for usage.
Estimating Procedure. Using the plans, count each fixture.
Example. An example is not needed.
Accuracy Rate. The accuracy rate should be 100 percent.
Scrap Consideration. This consideration is not applicable.

Appliances

Description. A determination must be made as to whether the appliances are gas or electric. You might want to include them as an allowance item.
Use. See the description for usage.
Estimating Procedure. Using the plans and kitchen elevations, determine which items are included and count them.
Example. An example is not needed.
Accuracy Rate. The accuracy rate should be 100 percent.
Scrap Consideration. This consideration is not applicable.

COMPLETING A TAKE-OFF FORM

The success of any job depends on the proper completion of an estimate form such as the one just described. Although Table D.5 is an estimate based on a preengineered home, most estimating techniques also apply to panelized and stick-built construction.

To summarize or recap the estimate, use Table D.5 as an sample take-off form and create one of your own to obtain the total cost of the job.

TABLE D.5 Material Take-Off Form

Foundation	Quantity	Options	Unit cost	Total cost
For reference: Total slab area	1180 sq. ft.			
Total grade beams	251 lf.			
A. Cushion sand	12 cu. yds.		7.22/yd.	86.64
B. Re-bar:				
1 #5s × 20'0"	50 pcs.		4.10 ea.	205.00
(Used at grade beams. Field bending at corners is required. Quantities listed allow for 12" laps at all splices.)				
2 #3s × 20'0"	88 pcs.		1.60 ea.	140.80
(Used at slab. Quantities listed allow for 12" laps at all splices. Quantity listed requires use of most of the drops.)				
Option: 6 × 6/10.10 welded-wire mesh in lieu of re-bar at slab. 6'-wide roll (with 6" laps)		164 lf.		
Note: If post tension cables are used in lieu of mesh or re-bar, consult local foundation engineer.				
3 #3s stirrups	168 pcs.		1.08 ea.	181.44
C. Vapor barrier (6 mil Visqueen)				
1 At entire slab area	1180 sq. ft.			27.42
2 Only at areas of slab to receive floor covering		1000 sq. ft.		
D. Anchor bolts—⅝" dia. × 12" lg. w/2" hook	28 pcs.		75¢ ea.	21.00
Option: Chemical anchors		TSS pkg		
E. Concrete—(3000 psi–5 sack)	33 cu. yds.		37.50 yd.	1237.50
Note: If 5½" ledge is used, add 0.0035 cubic yard concrete per lin. ft. of brick ledge required.				

TABLE D.5 Material Take-Off Form (Continued)

Exterior shell	Quantity	Options	Unit cost	Total cost
A. Roof				
1 30# felt × 36" wide × 72" long roll (installed with 6" side and endlaps)	10 rolls		7.97 roll	79.70
2 Shingles (type and color to be selected)	17 squares		20.95 sq.	398.05
3 Metal flashings				
a Drip edge—(10' sections)	170 lf.		1.02 ea.	17.34
b Valley flash—20" metal roll flashing 50' roll	72 lf.		28.44 ea.	56.88
c 6" metal roll flash	N/A			
d Tin shingles (step flashing) (used at chimney)	N/A			
4 Fasteners				
a Sheetmetal-washered nails × ¾" long (50 per square @ 100 per pound)	9 lbs.		89¢ lb.	8.01
b Shingle nails (to be determined by types of shingles used)				18.90
				578.88
B. Cornice (Defined as all soffit and fascia material, and all exterior wooden trims, including those above and below windows.)				
1 Soffit—4'0" × 8'0" (1 cuts 2) (materials can be ⁷⁄₁₆" masonite, ¼" or ⅜" plywood, smooth or rough-sawn finish)	12 pcs.		7.69 sheet	92.28
2 Fascia—8"-wide material (Material can be ⁷⁄₁₆" masonite or 1-×-8 lumber.)	170 lf.		38¢ lf.	66.88
3 Shingle mold (Material can be 1-×-2, 1-×-3, or 1-×-4 finger-jointed wood.)	170 lf.		24¢ lf.	42.24
4 Batten strips at soffit joints (¼" × ¾" screen molding)	56 lf		11¢ lf.	6.16
5 Soffit to siding trim				
a Brick frieze 2 × 4	88 lf.		2.67 ea.	24.03
b Trim (Material can be 1-×-3, or 1-×-4 finger-jointed wood.)	190 lf.		24¢ lf.	46.08
6 Window and door trims (Types of material are determined by types of windows and siding used.)				
a Window jambs—1 × 6 × 12'0"	12 lf.			3.09
b Window headers	N/A			
c Window sills	N/A			
d Door jambs	N/A			

TABLE D.5   Material Take-Off Form (Continued)

Exterior shell	Quantity	Options	Unit cost	Total cost
e Door headers	N/A			
f Garage door jamb & header facing 1" × 10" × 8'0"	22 lf.		2.04 ea.	6.12
g Garage door stop molding (pattern to be selected)	22 lf.		36¢ lf.	8.64
h Garage door jamb trim	14 lf.		36¢ lf.	5.76
i Garage door header trim	8 lf.		36¢ lf.	2.88
j Corner trim	N/A			
7 Porch soffit				
a Soffit 4'0"-×-8'0" sheets (Materials can be ⅜₆" masonite ¼" × ⅜" plywood, smooth or rough-sawn finish.)	N/A			
b Soffit batten strips (¼" × ¾"screen mold)	N/A			
c Soffit to siding trim (Material can be 1-×-3 or 1-×-4 finger-jointed wood.)	N/A			
8 Wooden nailer 2" × 10"	N/A			
9 Garage door framing 2" × 4" × 8'0" LG #2	22 lf.		1.35 ea.	4.05
				308.21
C. Siding				
Total square footage of sides of house	1552 sq. ft.			
Deduct for windows and doors shown on plans (including siding material around doors and windows.)	(–) 191 sq. ft.			
Deduct for gable area.	(–) 347 sq. ft.			
1 Synthetic plaster (Is a generic term for an insulated plaster or stucco installation system and generally is purchased installed as a subcontract.)	N/A			
2 Brick 2⅔ × 8 STD Modular (Area shown on plans) (window and door area deducted) (Actual brick count is dependent on brick size used.) 7155 Req'd	1060 sq. ft.		198.50 per thousand	1420.27
3 Siding (Material to be selected) (Window and door area deducted, gable area inclu⌐ːd)				
a 4"-wide siding		1092 lf.		
b 6"-wide siding × 8' 0" LG 93 pcs	744 lf.		19.36 ea.	1800.48
c 8"-wide siding		570 lf.		
				3220.75

TABLE D.5 Material Take-Off Form (Continued)

Exterior shell	Quantity	Options	Unit cost	Total cost
D. Windows				
(style and material to be selected)				
1 2630 single-hung, double-glazed, with half screen	1		49.88 ea.	49.88
2 3050 single-hung, double-glazed, with half screen	7		73.25 ea.	512.75
NOTE: If sizes or quantities of windows are changed, an adjustment has to be made in the quantity of siding and trim listed elsewhere.				
E. Miscellaneous exterior materials				
1 Lintels				
a Angle 3" × 3" × ⁵⁄₁₆" × 3'0"	2		8.88 ea.	17.76
b Angle 3" × 3" × ⁵⁄₁₆" × 3'6"	7		10.36 ea.	72.52
c Angle 4" × 4" × ⁵⁄₁₆" × 8'8"	1		17.10 ea.	17.10
2 Gable vents				
a Pitch, galv. fixed-blade vent	N/A			
b Pitch, galv. fixed-blade vent	N/A			
c 36" diameter, galv. fixed-blade vent	3		93.80 ea.	281.40
d diameter, galv. fixed-blade vent	N/A			
e Architectural vent: refer to drawings	N/A			
3 12"-diameter, fluted decorative column with cap and base × 7'10" lg. approximately (field measure required)	N/A			
4 Brick flash—6 mil Visqueen × 8" wide	136 lf.			28.50
5 Column wrap—1" × 12" (material to be selected) (Brick columns are listed on pg. 3)	N/A			
6 Skylights (size and type to be selected)	N/A			
7 Cricket (size depends on chimney width)	N/A			
8 Decorative shutters—12" × 5'0"	4 ea.		17.00 ea.	68.00
				485.28
F. Wood				
1 Roof wood #4 × 8	55			
2 Wall sheathing #4 × 8	48			
3 Floor decking a #8-18 × 1¹⁵⁄₁₆" screw	N/A			
G. Insulation				
1 Exterior insulation	992 sq. ft.			

TABLE D.5   Material Take-Off Form (Continued)

Exterior shell	Quantity	Options	Unit cost	Total cost
2 Attic insulation	908 sq. ft.			

Interior	Quantity	Options	Unit cost	Total cost
A. Drywall				
1 Gypsum wallboard (4' × 8' × ½") (Estimating formula: sq. ft. floor of house to receive gypsum wallboard divided by 10; plus sq. ft. of vaulted portion of walls divided by 32)	117 pcs.		2.28 sheet	263.25
Option: Some codes require 1-hour fire-separation between garage and house, and require ⅝"-thick fire-rated gypsum wallboard. Therefore, the quantities are:				
4' × 8' × ½" regular gypsum wallboard		105 pcs.		
4' × 8' × ⅝" fire-rated gypsum wallboard (Caution: Construction techniques must be approved by local fire marshall.)		12 pcs.		
2 Ready-mix joint compound (pre-mixed) 5 gal. pail (Estimating formula: 5.75 gals. per 1000 sq. ft. of surface area. Application techniques and job conditions cause this amount to vary)	22 gals.		7.33 pail	36.65
Option: If walls are to be textured, add approximately 4 gals. per 1000 sq. ft. of wall area.		15 gals.	7.33 pail	21.99
3 Joint tape—250' rolls (Estimating formula: 370 lin. ft. per 1000 sq. ft. of surface area)	6 rolls		1.75 roll	10.50
4 Drywall screws—type-S point, 39.84 buglehead × 1¼" lg. (Estimating formula: industry standard is one screw per 1 sq. ft., however, Tri-Steel Structures ceiling furring strips are 16" O.C. therefore, assume 1065 screws per 1 sq. ft. @ 258 pcs. per lb.)	16 lbs.		2.49 lb.	

TABLE D.5 Material Take-Off Form (Continued)

Interior	Quantity	Options	Unit cost	Total cost
5 Metal corner beads— 1¼ × 1¼ × 8'0" lg.	25 pcs.		78¢ ea.	19.50 391.73
B. Miscellaneous framing				
1 Furnace platform				
a 2" × 4" × 8'0" framing	3 pcs.		1.35 ea.	4.05
b 4' × 8' × ¾" CDX plywood	1 pc.		11.48 ea.	11.48
2 Attic access				
a 2" × 4" × 8'0" framing	3 pcs.		1.35 ea.	4.05
b Molded trim	10 lf.		42¢ ea.	4.20
c Pull-down attic stair	N/A			23.78
3 Prefabricated metal fireplace (complete with flute) (size & style to be selected)	N/A			
C. Trims				
1 Base wall moulding 3¼" wide × 12'0" lg. (Material size and pattern to be selected)	241 lf.		53¢ lf.	127.73
Option: If base wall moulding is used at garage, add		55 lf.		
2 1 × 10 window stool (1½" revel allowed at each side.) (Type of wood and edge pattern to be selected) NOTE: Window stool cannot be spliced, refer to drawings for exact lengths.	26 lf.		2.46 lf.	78.72
3 Window stool trim 2½" wide (Material size and pattern to be selected)	26 lf.		42¢ lf.	13.44
4 Shelving				
a Utility (closets, etc.)				
1 12" wide 1 × 12 × 8'0" (material to be selected)	32 lf.		6.72 ea.	26.88
2 16" wide (material to be selected)	N/A			
3 18" wide (material to be selected)	N/A			
b Decorative/adjustable (wet bar, etc.)				
1 12" wide (material to be selected)	N/A			

TABLE D.5   Material Take-Off Form (Continued)

Interior	Quantity	Options	Unit cost	Total cost
NOTE: Shelving cannot be spliced, refer to drawings for exact lengths.				
c 1 × 3 Utility shelving support × 8'0" lg	41 lf.	36¢ lf.		17.28
NOTE: Refer to hardware section for decorative shelving brackets.				
5 1¼" diameter closet rods × 14' lg	27 lf.		47¢ lf.	13.16
NOTE: Closet rods cannot be spliced, refer to drawings for exact lengths.				277.21

Interior	Sch. #	Quantity	Options	Unit cost	Total cost
D. Doors					
1 Entry doors (style and wood type to be selected)					
a 3068 patterned-face, insulated metal, prehung, with threshold	1	1			117.00
b 2868 flush-face, insulated metal, prehung, with threshold	2	1			112.00
c 2668 wooden or metal french, prehung with threshold	10	1			184.00
d 8'0" × 7'0" sectional overhead garage door with track and hardware	9	1			204.00
e					
					617.00
NOTE: 1. Some codes require special considerations for entry doors; refer to door schedule on drawings for notes.					
2. Above doors require 9"-wide jambs.					
2 Interior passage doors (style and wood type to be selected)					
a 2868 flush-panel, hollow-core wood, prehung	3	3		31.66 ea.	94.98

TABLE D.5 Material Take-Off Form (Continued)

Interior	Sch. #	Quantity	Options	Unit cost	Total cost
b 2068 flush-panel, hollow-core wood, prehung	4 & 5	4		27.66 ea.	110.64
c 3068 flush-panel, hollow-core wood, bifold, prehung	6	1			24.00
d 4068 twin, flush-panel, hollow-core wood, bifold, prehung	7	1			38.00
e 6068 twin, flush-panel, hollow-core wood, bifold, prehung	8	1			44.00
					311.62

Interior	Quantity	Options	Unit cost	Total cost
E. Floor covering (Refer to Room Finish Schedule on drawings for location of various floor coverings.)				
1 Carpet w/pad, 12'0"-wide roll (Color and style to be selected, amount can vary depending on pattern repeat.)	79 yds.		12.00 yd.	948.00
Option: Stairs carpeted		N/A		
2 Vinyl tile 12" × 12" × ¾₂" VCT (color and style to be selected)	81 sq. ft.		6.00 yd.	54.00
				1002.00
3 Sheet vinyl, 6'0"-wide roll (Color and style to be selected, amount can vary depending on pattern repeat.)	N/A			
4 Ceramic tile (Color and style to be selected) Caution: Quarry tile requires a recess in the floor slab, and if used, style should be picked before foundation work begins.	N/A			
F Wall cover				
1 Wallpaper (Quantity shown is sq. ft. of wall surface to be covered. Roll width and pattern repeat determines number of rolls required.)	N/A			
2 Ceramic tile (at showers and tubs) (color and style to be selected)				
a Tile				
1 70" AFF	44 sq. ft.		4.50 sq. ft.	198.00
2 Full height	65 sq. ft.			

TABLE D.5   Material Take-Off Form (Continued)

Interior	Quantity	Options	Unit cost	Total cost
b Ceramic soap dish	1		5.97 ea.	5.97
c 24" ceramic towel bar	1		11.97 ea.	11.97
				215.94
G Millwork				
1 Special architectural wooden stairs (design, material, and finish to be selected)	N/A			
2 Cabinets (Quantities shown are linear feet) (style, material, and finish to be selected)				
a Upper cabinets	20'0"		19.00 lf.	380.00
b Base cabinets	10'9"		20.00 lf.	215.00
c Island with cook top	N/A			
d Cabinet face (linen front)	N/A			
				595.00
NOTE: Refer to drawings for cabinet details.				
H Tops (Quantities shown are linear feet) (style, material, and finish to be selected)				
1 Vanity top × 1'9" wide with 4"-backsplash	4'3"		9.75 lf.	41.44
2 Plastic laminated countertop × 2'0" wide with 4"-backsplash	20'8"		9.75 lf.	202.31
3 Plastic laminated countertop × wide	N/A			
4 Plastic laminated bar top × wide	N/A			
5 Plastic laminated bar top × wide	N/A			
				243.75
I Glass and Mirrors				
1 Glass shower door—custom wide shower front with 2'0" wide door	N/A			
2 Mirrors				
a ¼" tempered mirror, size 48" × 42"	1			59.50
b ¼" tempered mirror, size				
c ¼" tempered mirror, size				
d ¼" tempered mirror, size				
e ¼" tempered mirror, size				
NOTE: Field measurment required				
J Hardware (materials, styles, and colors to be selected) 1 Bathroom				

TABLE D.5 Material Take-Off Form (Continued)

Interior	Quantity	Options	Unit cost	Total cost
a Towel bar	1			5.09
b Paper holder	1			6.98
c Medicine cabinet–recessed	1			9.99
d Medicine cabinet–surface mounted	N/A			
e Shower curtain rod	1			8.50
2 Closet and shelves				
a 1¼" diameter rod hanger bracket–center	3		1.59 ea.	4.77
b 1¼" diameter rod hanger bracket–end	8		69¢ ea.	5.52
c 12" fixed-shelf brackets (at closets)	2		1.89 ea.	3.78
3 Cabinets				
a Door knobs	22		1.50 ea.	33.00
b Drawer pulls	7		1.50 ea.	10.50
c Cabinet door catches	22		40¢ ea.	8.80
4 Doors				
a Main entry lockset	1			55.00
b Rear entry lockset	1			13.75
c Atrium door lockset	N/A			
d Garage/kitchen door lockset	1			13.75
e Privacy lockset	5		8.98 ea.	44.90
f Passage latch set	N/A			
g Closet latch set	2		7.95 ea.	15.90
h Bifold door pulls	5		3.00 ea.	15.00
i Bullet catches	5		99¢ ea.	4.95
j Door stops	8		2.00 ea.	16.00
k Garage door opener	1			139.00
l Door closure		2		
NOTE: 1. It is customary to key all exterior doors alike.				415.18
2. Some codes require special consideration for exterior locksets and hinges. Refer to door schedule on drawings for notes.				
K Plumbing fixtures (manufacturer's model number or style, and color to be selected) NOTE: These items typically are part of the plumbing subcontract price.				
1 Hot water heater (50 gal. gas)	1			249.00
2 Double kitchen sink with disposal	1	sink		59.00
		disposal		76.00
3 Single bar sink with gooseneck spigot	N/A			
4 Toilet	1			80.00
5 Lavatory	1			49.50
6 Bath tub	1			228.00
				741.50

TABLE D.5 Material Take-Off Form (Continued)

Interior	Quantity	Options	Unit cost	Total cost
L Appliances (manufacturer model or style, and color to be selected)				
1 Dishwasher–under counter	1			325.00
2 Oven–slid-in type	N/A			
3 Microwave oven–slid-in type	N/A			
4 Cook surface with grill	N/A			
5 4-burner cook surface	N/A			
6 Range with oven (gas)	1			325.00
7 Range hood				
a Vented	1			35.00
b Ventless	N/A			
				685.00

Estimate re-cap

Item	From page no.	Material	Labor	Sub-contract	Total
Foundation					
A Cushion sand	1	86.64			86.64
B Re-bar	1	527.24			527.24
C Vapor barrier	1	27.42			27.42
D Anchor bolts	1	21.00			21.00
E Concrete	1	1237.50			1237.50
Labor			1227.20		1227.20
Exterior shell					
A Roof	2	578.88	284.86		863.74
B Cornice	2 & 3	308.21	400.00		708.21
C Siding	3 & 4	1800.48	372.00		2172.48
(brick)	3	1420.27	1440.00		2860.27
D Windows	4	562.63	(in erection)		562.63
E Misc. exterior material	4 & 5	485.28			485.28
F Exterior wood					
G Insulation					
Interior					
A Drywall	6	391.73	866.53		1258.26
B Misc. framing	6 & 7	23.78	75.00		98.78
C Trims	7	277.21	(in d)		277.21
D Doors	8	928.62	590.00		1518.62
E Floor covering	9			1002.00	1002.00
F Wall covering	9	215.94	250.00		465.94
G Millwork	10	595.00	300.00		895.00
H Tops	10	243.75	200.00		443.75
I Glass & mirrors	10	59.50	25.00		84.50
J Hardware	11	415.18	150.00		565.18
K Plumbing fixtures	12	741.50			741.50
L Appliances	12	685.00			685.00

TABLE D.5 Material Take-Off Form (Continued)

	Estimate re-cap				
Item	From page no.	Material	Labor	Sub-contract	Total
Steel erection				3107.88	3107.88
Basic home package	13	16,170.00			16,170.00
Electrical fixtures		750.00			750.00
Subcontract—Plumbing				1500.00	1500.00
Subcontract—HV/AC				2500.00	2500.00
Subcontract—Electrical				1500.00	1500.00
Subcontract—Painting				1164.00	1164.00
Final clean-up			500.00		500.00
	Total cost	28,552.76	6,680.59	10,773.88	46,007.23

APPENDIX E
SUGGESTED SPECIFICATIONS

The following suggested specifications are intended to serve as general guidelines. They must be modified to suit the conditions of use, with particular attention given to the deletion of inapplicable provisions. An italicized comment, where shown, serves as a commentary to the item it follows.

1.0 GENERAL

1.1 Description of Work

Furnish all labor, materials, services, and equipment necessary to complete all cold-formed steel framing work shown on the drawings and/or as specified herein.

1.2 Related Work Specified Elsewhere

Specifier should list applicable specification references for items that must be coordinated with the installation of the framing system.

1.3 Submittals

The following items shall be furnished by the contractor prior to fabrication.
1.3.1 Shop Drawings. Drawings for review that include:

- Cross sections, plans and/or elevations depicting component types and locations for each unique framing application
- Connection details depicting fastener type, quantity, location, and other information to assure proper installation

Contractors electing to install prefabricated/prefinished frames, shall submit drawings depicting panel configurations, dimensions, components, locations, and construction sequence.

1.3.2 Structural Calculations. These depend on design criteria, structural analysis for each unique framing application, and selection of framing components and accessories and verification of connections. This item can be excluded if the contract documents include the results of an analysis performed by the owner's engineer. Should calculations be required, the owner's architect or engineer shall develop the design criteria. This should include maximum allowable deflections, wind velocity in miles per hour, applicable model building code, gravity and live loads, etc.

1.3.3 Certifications. Submit statements from the framing manufacturer certifying conformance with applicable standards outlined in Part 2.3.

1.3.4 Literature. Submit technical literature prepared by the framing manufacturer.

1.4 Quality Assurance

1.4.1 Standards. Products and installation shall meet the requirements of the following standards, unless more stringent requirements are described herein:

- American Iron and Steel Institute's (AISI) *Specification for the Design of Cold Formed Steel Structural Members*, 1986 edition and 1989 addendum
- American Welding Society's (AWS) *Specification for Welding Sheet Steel in Structures*, D1.3
- American Society for Testing and Materials' (ASTM) specifications:
  ~A-446 Standard Specification for Sheet Steel, Zinc Coated by the Hot Dip Process, Structural (Physical) Quality C-955 Standard Specification for Load-Bearing (Transverse and Axial) Steel Studs, Runner (Track) and Bracing and Bridging, for Screw Application of Gypsum Board and Metal Plaster Bases
  ~C-1007 Standard Specification for the Installation of Load-Bearing (Transverse and Axial) Steel Studs and Related Accessories

1.4.2 Inspection and Quality control. Inspections shall be performed by an independent authority to check conformance with the provisions of the contract documents and approved erection drawings. The contractor assumes all responsibility for quality control for the scope of work described in 1.1.

2.0 PRODUCTS

2.1 Acceptable Manufacturers (as Stated in Contract)

2.2 Structural Properties

Framing products shall possess the minimum structural properties listed below (refer to List of Acronyms for definitions):

								Net return	
Member	Area	Web	Flange	Lip	Ix-x	Sx-x	Rx-x	Fy	MRes
	(in^2)	(in)	(in)	(in)	(in^4)	(in^3)	(in)	(KSI)	(K-in)

When the owner's engineer of record has performed the structural analysis as discussed in item 1.3.2, prepare this table of minimum physical and structural properties.

2.3 Stud Type

All stud (and/or) joist framing members shall be of the type, size and gauge as shown on the plans.

2.4 Material

2.4.1 ASTM A570 and ASTM A446. All painted studs and joists of 12-, 14-, and 16-gauge shall be formed from steel that corresponds to ASTM A570. Galvanized studs and joists of 12-, 14-, and 16-gauge shall be formed from steel that corresponds to ASTM A446. Structural calculations should be prepared utilizing one of the following grades. (Catalog charts are based on grade D.)

• Grade A minimum yield strength 33

• Grade C minimum yield strength 40

• Grade D minimum yield strength 50

2.4.2 ASTM A611, Grade C. All painted 18- and 20-gauge studs and/or joists, and all painted track, bridging, end closures, and accessories shall be formed from steel that corresponds to the requirements of ASTM A611, grade C, with a minimum yield of 33,000 psi.

2.4.3 ASTM A446, Grade A. All galvanized 18- and 20-gauge studs and/or joists, and all galvanized track, bridging, end closures, and accessories shall be formed from steel that corresponds to the requirements of ASTM A446, grade A, with a minimum yield of 33,000 psi.

2.4.4 ASTM A525 and C955. Manufacturers suggest all studs, joists and accessories be formed from steel having a G-60 galvanized coating, meeting ASTM A525 and C955.

2.4.5 Physical and Structural Properties. The physical and structural properties listed by the manufacturers shall be considered the minimum permitted for all framing members. Specifically, the following minimum properties, calculated in accordance with the latest AISI specifications, shall be provided (1986):

Component	Ix	Resisting moment
(Stud, joist or accessory)	$(inch_4)$	(inch/pound)

3.0 EXECUTION

3.1 Storage of Materials

Products shall be protected from conditions that might cause any physical damage. Any damaged material shall be removed from the job site.

3.2 Installation: General

Methods of construction can be either piece-by-piece, or stick-built, or by fabrication into panels either on or off the site.

3.2.1 Connections. Connections shall be accomplished with self-drilling screws or welding so that the connection meets or exceeds the design loads required at that connection.

3.2.2 Studs. Studs that are only loaded transversely need not sit squarely and butt the track web, but must be attached to both track legs.

3.2.3 Axial-Loaded Studs. These members shall be installed so that they seat squarely, within $\frac{1}{16}$ inch, against the web portion of the top and bottom tracks. Tracks shall rest on a continuous uniform bearing surface. Studs must be attached to both track legs.

3.2.4 Cutting. Steel-framing members can be cut with a saw, shear, or plasma cutter. Torch cutting of load-bearing members is not permitted.

3.2.5 Temporary Bracing. Leave temporary bracing in place until work is stabilized.

3.2.6 Bridging. Bridging shall be of a size and type shown on the drawings and as called for in the design calculations and working drawings.

3.2.7 Headers. Install headers in all openings in axial-loaded walls that are wider than the stud spacing in that wall, and formed as shown on the shop drawings.

3.2.8 Insulation. Insulation equal to the job requirements shall be placed in all jambs, headers, and corner conditions that will be inaccessible after their installation into the wall.

3.2.9 Jackstuds. Jackstuds that provide support at each end of the headers and that are connected securely to the header, must seat squarely in the lower and upper track of the wall and header, and must be properly attached to them.

3.2.10 Wall Track. Do not use wall track to support any load.

3.2.11 Alignment of Axial-Loaded Members. All axial-loaded members shall be aligned vertically to allow for the full transfer of the loads down to the foundation. Vertical alignment shall be maintained at floor/wall intersections or alternate provisions for load transfer must be made.

3.2.12 Field-Cut Holes. Holes that are field cut into steel framing members shall be within the limitations of the product and its design. Provide reinforcement when holes are cut through load-bearing members in accordance with the manufacturer's recommendations and as approved by the project architect or engineer of record.

3.2.13 Finishing Steel. All steel that has been cut or welded shall be touched up with zinc-rich paint.

3.2.14 Spacing Studs. Studs shall be spaced to suit the design requirements and limitations of collateral facing materials.

3.2.15 Attaching Gypsum Board. Gypsum board shall be attached to steel studs in accordance with ASTM specification C-840.

3.2.16 Providing for Structural Movement. Provision for structural movement shall be allowed where indicated and necessary by design or code requirements.

3.2.17 Prohibited Splicing. Splicing of axial-loaded members shall not be permitted.

4.0 ERECTION OF AXIAL-LOAD-BEARING MEMBERS

4.1 Prefabricated Panels

Handling and lifting of prefabricated framing panels shall be done in a manner that does not cause distortion in any member.

4.2 Tracks

Tracks shall be securely anchored to the supporting structure as shown on the plans.

4.2.1 Bearing Supports. Complete uniform and level bearing support shall be provided for the bottom track.

4.2.2 Abutting Joints. At track butt joints, abutting pieces of track shall be securely anchored to a common structural element, or they shall be butt welded or spliced together.

4.2.3 Stud Alignment. Studs shall be plumbed, aligned, and securely attached to the flanges or webs of both the upper and lower tracks.

4.2.4 Wall Openings. Framed wall openings shall include headers and supporting studs as shown on the plans.

4.2.5 Jackstuds. Jackstuds shall be installed below window sills, above window and door heads, at freestanding stair rails, and elsewhere to furnish support, and shall be securely attached to supporting members.

4.2.6 Temporary Bracing. Temporary bracing shall be provided until erection is completed.

4.2.7 Wall-Stud Bracing. Wall-stud bridging shall be installed in a manner that provides resistance to both minor axis bending and rotation. Bridging rows shall be equally spaced not to exceed 5-foot centers for wind loading only, or 4-foot centers for axial loading.

4.2.8 Stud Walls. Provide stud walls at locations indicated on plans as shear walls for frame stability and lateral-load resistance. Such stud walls shall be braced as indicated on plans and specifications. Additional studs shall be positioned to resist the vertical components as indicated on plans.

4.2.9 Splices. Splices in axial-loaded studs shall not be permitted.

4.2.10 Insulation. Provide insulation equal to that specified elsewhere in all doubled header members that will not be accessible to the insulation contractor.

5.0 ERECTION OF COLD-FORMED STEEL JOISTS

5.1 Joist Location

Joists shall be located directly over bearing studs or a load-distributing member shall be provided at the top track.

5.2 Web Stiffeners

Provide web stiffeners at reaction points where indicated by plans.

5.2.1 Joist Bridging. Joist bridging shall be provided as shown on the plans.

5.2.2 Additional Joists. Provide an additional joist under parallel partitions when the partition length exceeds one-half the joist span and around all floor and roof openings that interrupt one or more spanning members unless otherwise noted.

5.2.3 End Blocking. End blocking shall be provided where joist ends are not otherwise restrained from rotation.

6.0 INSTALLATION: GENERAL

These items should be reviewed and edited for the application in question.

6.1 Prefabricated Frames

Prefabricated frames shall be square, with components attached in a manner to prevent racking during fabrication, transportation, and lifting. Provisions to lift the panel shall be included in the frame's design and construction.

6.2 Field-Cuts

Field cutting of steel framing members shall be by saw or plasma cutter. Torch cutting is not permitted except by written approval of the engineer of record.

6.3 Removal of Temporary Bracing

Temporary bracing shall be provided and remain in place until work is permanently stabilized.

6.4 Insulation

Insulation, equal to the specified requirements shall be placed in components inaccessible to the insulation contractor after their installation.

6.5 Splicing

When the splicing of track is necessary between stud spacings, a piece of stud shall be placed between adjacent tracks and fastened by weld or screw to each side of the track and each end. Splicing of framing components, other than track, is not permitted.

6.6 Stud Spacing

Studs shall be spaced as shown (in the contract documents or approved shop drawings) or as required to meet the design requirements and limitations of the collateral materials.

6.7 Sealants

A sealant shall be applied to concrete or masonry surfaces prior to anchoring tracks.

7.0 INSTALLATION: NONLOAD-BEARING (CURTAIN) WALLS

7.1 Studs

Studs shall be plumbed, aligned, and secured to the continuous runner tracks at each end and each side, unless the stud end terminates at a deflection track.

7.2 Mechanical Bridging

Mechanical bridging, of the type and spacing described (in the contract documents or approved shop drawings), shall be installed prior to the installation of facing materials.

7.3 Collateral Materials

Installation of sheathing, wallboard, or any other collateral material shall be performed in accordance with the product manufacturer's specifications.

7.4 Component Locations

Components, e.g., deflection track and/or slide clips, shall be provided at locations described in the contract documents to accommodate potential primary frame movements. Construction shall accommodate a vertical displacement of ___ inch. *Input the specified deflection.*

7.5 Jambs

Multiple studs shall be provided at the jambs of wall openings. Individual studs forming the jamb shall be attached together at intervals described in the contract documents or approved shop drawings.

8.0 *INSTALLATION: AXIAL-LOAD BEARING WALLS*

8.1 Studs

Studs shall be installed seated squarely, within $\frac{1}{16}$ inch, against the web of the top and bottom track to assure the transfer of axial load. Studs shall be plumbed, aligned, and secured to the continuous runner tracks at each end and each side before the installation of components that induce axial load.

8.2 Track

Track shall rest on a continuous bearing surface. If not provided, install full-size shims below track at stud locations or set bottom track in high-strength grout.

8.3 Bridging

Bridging, of the type and spacing described in the contract documents or approved shop drawings, shall be installed before loading.

8.4 Framed Wall Openings

Framed wall openings shall include headers and supporting components as shown in the contract documents or approved shop drawings.

8.5 X Bracing Assemblies

Installation of shear-wall X bracing assemblies, as shown in the contract documents or approved shop drawings, shall be completed before the attachment of facing materials and/or the erection of ascending levels.

8.6 Distribution Members

Where the floor or roof components do not directly align over a stud, a continuous distribution member shall be provided at the top of the wall. Do not use the top track as a distribution header.

9.0 INSTALLATION: HORIZONTAL JOISTS OR RAFTERS

9.1 Web Stiffeners

Provide web stiffeners at support locations where indicated in the contract documents or approved shop drawings.

9.2 Mechanical Bridging

Mechanical bridging, of the type and spacing described in the contract documents or approved shop drawings, shall be installed before loading.

9.3 Additional Joists

Provide an additional joist under parallel, nonload-bearing partitions when the partition length exceeds one-half the joist span.

9.4 End Blocking

End blocking shall be provided where joist ends are not restrained against rotation.

9.5 Framing Openings

Provide additional framing around openings, as shown in the contract documents or approved shop drawings, when the width of the opening exceeds the typical joist/rafter spacing.

9.6 Floor/Rafter Systems

During construction, the floor/rafter system shall not be loaded beyond the limits for which it was designed.

10.0 INSTALLATION: TRUSSES

10.1 Trusses

Trusses shall be bridges, as shown in the contract documents or approved shop drawings, before the installation of collateral materials.

10.2 Temporary Bracing

Temporary bracing shall be provided and remain in place until work is permanently stabilized.

11.0 CONNECTIONS

11.1 Welds

Welds shall be of the type, size, and location shown in the contract documents or approved shop drawings. Welded connections shall be performed in accordance with the American Welding Society's (AWS) *Specification for Welding Sheet Steel in Structures*, D1.3. Welders, welding operations, and welding procedures shall be qualified in accordance with AWS D1.3. *Consult applicable AWS specifications for information regarding safe welding procedures.* Welds shall be cleaned and coated with rust inhibitive galvanizing paint.

11.2 Screws

Screws shall be of the type, size, and location shown in the contract documents or approved shop drawings. Screw penetration through joined materials shall not be less than three exposed screw threads.

    11.2.1 Protective Coatings. Specifier should contact screw manufacturers for information regarding protective coatings and include that information herein.

    11.2.2 Installation Instructions. Contractor shall refer to installation instructions published by the screw manufacturer and ASTM C-954 for minimum spacing and edge-distance and torque requirements.

11.3 Concrete Anchors

    11.3.1 Types. Anchor bolts, epoxy bolts, wedge expansion bolts, screw-type concrete fasteners, and powder actuated fasteners can be used as concrete anchors.

    11.3.2 Shear and Tension Capacities. The shear and tension capacities of the fasteners must be verified for the application in question. Bearing capacity of the supported element should be checked in accordance with the AISI specification.

11.3.3 Installation. Concrete anchors shall not be installed until full compressive strength is obtained.

11.3.4 Installation Instructions. Contractor shall refer to instructions published by the anchor manufacturer for minimum spacing, edge distance, concrete embedment, and additional installation requirements.

11.4 Substitutions

The contractor can substitute fasteners of equivalent specifications and load carrying capacities.

11.5 Wired-Tied Connections

Wired-tied connections in structural applications shall not be permitted.

12.0 TOLERANCES

12.1 Vertical Alignment

Vertical alignment of studs shall be within L/960, which is ⅛ inch in 10 feet, of the span.

12.2 Horizontal Alignment

Horizontal alignment of walls shall be within L/960, which is ⅛ inch in 10 feet, of their respective lengths.

12.3 Stud Spacing

Spacing of studs shall not be more than ⅛ inch from the designed spacing providing so that the cumulative error does not exceed the requirements of the finishing materials.

12.4 Prefabricated Panels

Prefabricated panels shall not be more than ⅛ inch out of square within the length of that panel.

13.0 INSPECTIONS

13.1 Purpose

Inspections shall be performed in order to assure strict conformance to the shop drawings at all phases of construction.

13.2 Check Members

All members shall be checked for proper alignment, bearing, completeness of attachments, reinforcement, etc.

13.3 Check Attachments

All attachments shall be checked for conformance with the shop drawings. All welds shall be touched up.

13.4 General Inspection

A general inspection of the structure shall be completed prior to applying loads to members.

GLOSSARY

acoustics science dealing with the production, control, transmission, reception, and effects of sound, and the process of hearing.

adhesion the ability of the membrane to remain adhered during its service life to the substrate or to itself.

adhesive a compound, glue, or mastic used in the application of gypsum board to framing or for laminating one or more layers of gypsum boards.

AFF above finish floor.

aggregate sand, gravel, crushed stone, or other material that is a main constituent of portland-cement concrete and aggregated gypsum plaster. Also polystyrene, perlite, and vermiculite particles used in texture finishes.

airborne sound sound traveling through the medium of air.

anchor metal securing device embedded or driven into masonry, concrete, steel, or wood.

anchor bolt heavy, threaded bolt embedded in the foundation to secure sill to foundation wall or bottom plate of exterior wall to concrete floor slab.

annular ring nail a deformed shank nail with improved holding qualities specially designed for use with gypsum board.

APA American Plywood Association.

apron the piece of window trim applied vertically beneath the sill or a hard-surfaced area in front of a building, such as the concrete apron of a garage.

area-separation wall residential fire walls, usually with a 2- to 4-hour rating, designed to prevent the spread of fire from an adjoining occupancy; extends from foundation to or through the roof. Identified by codes as either *fire wall, party wall,* or *townhouse separation wall.*

ASTM formerly American Society for Testing and Materials, now ASTM, a nonprofit, national technical society that publishes definitions, standards, test methods, recommended installation practices, and specifications for materials.

attenuation reduction in sound level.

balloon frame method of framing outside walls in which studs extend the full length or height of the wall.

balusters spindles that help support a staircase handrail.

bar joist   open-web, flat-truss structural member used to support floor or roof structure. Web section is made from bar or rod stock, and chords are usually fabricated from T or angle sections.

base plate   the lowest part of the steel structure. The welded plate on the end of a column that attaches to a foundation.

base track   the bottom track or plate of a metal stud wall.

batten   narrow strip of wood, plastic, metal, or gypsum board used to conceal an open joint.

bay   the interval of space created between frames.

beam   load-bearing member spanning a distance between supports.

bearing   support area upon which something rests, such as the point on bearing walls where the weight of the floor joist or roof rafter bears.

bearing partition   any interior divider that supports the weight of the structure above it. *Compare nonbearing partition.*

bearing wall   an interior or exterior wall that helps support the roof or the floor joists above.

bed   to set firmly and permanently in place.

bending   bowing of a member that results when a load or loads are applied laterally between supports.

blankets   fiberglass or rock-wool insulation that comes in long rolls 15 or 23 inches wide.

blocking   any nonstructural member in a floor, wall, or ceiling that serves primarily as a point for fastening finishing materials or accessories.

board-and-batten   a finished wall surface consisting of boards applied vertically with gaps between them and with battens covering the gaps.

board foot (Bd. Ft.)   volume of a piece of wood, 1 inch × 1 foot × 1 foot nominal. All lumber is sold by the board-foot measure.

brick flash   6-mil polyethylene film used at the base of brick veneer for waterproofing.

brick frieze   used to terminate the top of brick veneer at the soffit.

brick veneer   nonload-bearing brick facing applied to a wall to give the appearance of solid brick construction. Bricks are fastened to backup structure with metal ties embedded in mortar joints.

bridging   members attached between floor joists to distribute concentrated loads over more than one joist and to prevent the rotation of the joist. Solid bridging consists of joist-depth lumber installed perpendicular to and between the joists. Crossbridging consists of pairs of braces set in an X form between joists.

building codes   community ordinances governing the manner in which a home can be constructed or modified. Most codes primarily concern themselves with fire and health, with separate sections relating to electrical, plumbing, and structural work. Also see *Zoning.*

built-up roof   a surface for a flat or nearly flat roof that consists of several layers of roofing felt, each hot-mopped with asphalt, and a topping of gravel or crushed stone.

butt   to place materials end-to-end or end-to-edge without overlapping.

butt joint   joints formed by the mill-cut ends or by job cuts without a tapered edge. Synonym for *end joint.*

camber   curvature built into a beam or truss to compensate for loads that is encountered when in place and full dead load is applied. The crown is placed upward. Insufficient camber results in unwanted deflection when the member is loaded.

cant beam beam with edges chamfered or beveled.

cant strip triangular section laid at the intersection of two surfaces to ease or eliminate the effect of a sharp angle or projection.

cantilever any part of a structure that projects beyond its main support and is balanced on it. Also the act of projecting and balancing the structure.

carrying channel main supporting member of a suspended-ceiling system to which furring members or channels are attached.

casement glazed sash or frame hung to open like a door.

casing in foundations, a tube of wood or steel, usually circular, used to retain the walls of a deep, narrow excavation. In carpentry, the trim around windows, doors, columns, or piers.

caulk any of a variety of different compounds used to seal seams and joints against the infiltration of water and air.

ceiling beam the horizontal structural member that bolts at each end to a column and supports the ceiling furring.

cement board a factory-manufactured panel, ¼ to ¾ inch thick, 32 to 48 inches wide, and 3 inches to 10 feet long, made from aggregated and reinforced portland cement.

chalkline straight working line made by snapping a chalked cord stretched between two points, which transfers chalk to work surface.

choker a connecting cable used to lift frames and normally supplied by the crane company.

chord any principal member of a truss. In a roof truss the top chord replaces a rafter and the bottom chord replaces a ceiling joist.

cladding panels, bases, sheathing, cement board, etc., applied to framing.

clerestory a short (in height) exterior wall between two sections of roof that slope in different directions. The term also describes the window frequently used in such walls.

collar preformed flange placed over a vent pipe to seal the roof around the vent-pipe opening. Also called a vent sleeve.

collateral load this load is the weight of additional permanent materials such as mechanical and electrical systems, partitions, and ceiling. Collateral loads do not include dead loads.

column vertical, load-bearing member.

come-along a cabled, hand-operated winch used to adjust and square frames.

compression a force that presses particles of a body closer together.

compressive strength measures maximum unit resistance of a material to crushing load. Expressed as force per unit cross-sectional area, e.g., pounds per square inch (psi).

concrete footing generally wide, lower part of a foundation wall that spreads weight of building over a larger area. Its width and thickness vary according to the weight of building and type of soil on which building is erected.

condensation the process by which moisture in the air becomes droplets of water or ice on a surface whose temperature is colder than the air's temperature.

conduction, thermal transfer of heat from one part of a body to another part of that body, or to another body in close contact, without any movement of the bodies involved. The hot handle of a skillet is an example. Heat travels from the bottom of the skillet to the handle by conduction.

convection   the process of heat carried from one point to another by movement of a liquid or a gas (i.e., air). Natural convection is caused by the expansion of a liquid or gas when heated. Expansion reduces the density of the medium, which causes it to rise above the cooler, more dense portions of the medium.

coping   the covering course of a wall.

corner brace   structural framing member used to resist diagonal loads that cause racking of walls and panels due to wind and seismic forces. Can consist of a panel or diaphragm, or diagonal flat strap or rod. Bracing must function in both tension and compression. If brace only performs in tension, two diagonal tension members must be employed in opposing directions as X bracing.

corner post   a timber or other member forming the corner of a frame. Can be solid or built-up as a multipiece member.

cornice   all soffit and fascia material and all exterior wooden trims, including those above and below windows.

courses   parallel layers of building materials such as bricks, shingles, or siding laid up horizontally.

cricket   a superimposed construction placed in a roof area to assist drainage.

cripple   short stud such as that used between a door or window header and the top plate.

curtain wall   exterior wall of a building that is supported by the structure and carries no part of the vertical load except its own. Curtain walls must be designed to withstand and transfer wind loads to the structure.

cushion sand   fill material used to prepare foundation for concrete slab.

dead level   absolutely horizontal, of zero slope.

dead load   the load on a building element contributed by the weight of the building materials.

deflection   displacement that occurs when a load is applied to a member or assembly. The dead load of the member or assembly itself causes some deflection, as might occur in roofs or floors at midspan. Under applied wind loads, maximum deflection occurs at midheight in partitions and walls.

deformation   change in shape of a body brought about by the application of a force internal or external. Internal forces can result from temperature, humidity, or chemical changes. External forces from applied loads also can cause deformation.

design load   combination of weight (dead load) and other applied forces (live loads) for which a building or part of a building is designed. Based on the worst possible combination of loads.

dew point   the temperature at which air becomes saturated with moisture and below which condensation occurs.

door buck   structural element of a door opening. Can be the same element as the frame if frame is structural, as in the case of heavy steel frames.

door closure   hydraulic device mounted to the top of a door that automatically closes a door.

dormer   a structure that projects from a sloping roof, with at least one vertical wall large enough for a window or ventilator.

double-hung window   window sash that slides vertically and is offset in a double track.

downspout   a pipe for draining water from roof gutters. Also called a leader.

drip   an interruption or offset in an exterior horizontal surface, such as a soffit, immediately adjacent to the fascia. Designed to prevent the migration of water back along the surface.

drip edge   galvanized sheetmetal angle installed around the perimeter of the roof used to prevent rainwater from wicking back up under the roof shingles.

drywall generic term for interior surfacing material, such as gypsum panels, applied to framing using dry construction methods (e.g., mechanical fasteners or adhesive).

eave the part of a roof that projects beyond its supporting walls.

ell an extension of a building at right angles to its length.

endwall vertical wall constructed in line with the first or last frame of a structure.

expansion joint a structural separation between two building elements designed to minimize the effect of the stresses and movements of a building's components and to prevent those stresses from splitting or ridging the roof membrane.

exposure I grade type of plywood approved by the American Plywood Association for exterior use.

exterior insulation and finish system (EIFS) exterior cladding assembly consisting of a polymer finish over a reinforcement adhered to foam-plastic insulation that is fastened to masonry, concrete, building sheathing, or directly to the structural framing. The sheathing can be cement board or gypsum sheathing.

factor of safety ratio of the ultimate unit stress to the working or allowable stress.

fascia board board fastened to the ends of the rafters or joists forming part of a cornice.

fast track method that telescopes or overlaps traditional design-construction process. Overlapping phases as opposed to sequential phases is keynote of the concept.

fatigue condition of material under stress that has lost, to some degree, its power of resistance as a result of repeated applications of stress, particularly if stress reversals occur from positive and negative cyclical loading.

feathering blending of finishing coats along the edges to minimize ridges and sanding. Synonym for *cutting, wiping*.

felt fibrous material saturated with asphalt and used as an underlayment of sheathing paper.

fill-in stud the stud located between the point of support and floor (base) primarily found in curtain-wall construction.

fire resistance relative term, used with a numerical rating or modifying adjective to indicate the extent to which a material or structure resists the effect of fire.

fire-resistant refers to properties or designs that resist effects of any fire to which a material or structure might be expected to be subjected.

fire-retardant denotes substantially lower degree of fire resistance than fire-resistant. Often used to describe materials that are combustible, but have been treated to retard ignition or spread of fire under conditions for which they were designed.

fire stop obstruction in a cavity designed to resist the passage of flame, sometimes referred to as fire blocking.

fire taping the taping of gypsum-board joints without subsequent finishing coats. A treatment method used in attic, plenum, or mechanical areas where aesthetics are not important.

fire wall fire-resistant partition extending to or through the roof of a building to retard spread of fire.

flammable capability of a combustible material to ignite easily, burn intensely, or have rapid rate of flame spread.

flashing strips of metal or waterproof material used to make joints waterproof as in joining of curtain-wall panels.

flat spots areas on textured surfaces that have little or no aggregate. Synonym for *holidays*.

floor beam   the horizontal structural member that bolts at each end to a column and supports floor joists and ceiling furring.

fly rafter   the end rafter of a gable roof with a side overhang.

FM   Factory Mutual Research Corp.

footing   lower extremity of a foundation or load-bearing member that transmits load to load-bearing substrate.

foundation   a component that transfers weight of building and occupants to the earth.

frame   complete assembly of structural steel members containing rafters, ceiling beams, floor beams, etc.

frequency (sound)   number of complete vibrations, cycles, or periodic motion per unit of time.

furring   member or means of supporting a finished surfacing material away from the structural wall or framing. Used to level uneven or damaged surfaces, or to provide space between substrates. Also an element for mechanical or adhesive attachment of paneling.

gable dormer   dormer projection having a peaked roof with a triangle-shaped front.

gable roof   a roof shape characterized by two sections of roof of constant slope that meet at a ridge.

gambrel roof   a type of roof containing two sloping planes of different pitch on each side of the ridge. The lower plane has a steeper slope than the upper. Contains a gable at each end.

girder   a beam, especially a long, heavy one; the main beam supporting floor joists or other smaller beams.

glazing   the process of installing glass, which commonly is secured with glazier's points and glazing compound.

gusset   wooden or metal plate riveted, bolted, glued, or pressed (wooden trusses) over joints to transfer stresses between connected members.

gutter   a slightly sloping horizontal trough for catching water off a roof. Also called an eave trough.

header   horizontal framing member across the ends of the joists. Also the member over a door or window opening in a wall.

heat   form of energy thought to be characterized by the rate of vibration of the molecules of a substance. The hotter the substance, the faster the molecules vibrate. On the other hand, when there is no heat present it is thought the molecules are at rest, which theoretically occurs at absolute zero, -459.7 degrees F (-273.15 degrees C or 0.0 degrees K).

heat quantity (Btu)   common unit of measurement of the quantity of heat is the British Thermal Unit (Btu). One Btu is the amount of heat required to raise one pound of water from 63 degrees to 64 degrees F (1 Btu = 1055.06 J). This is about the amount of heat given off by one wooden match. A pound of coal can produce 13,000 Btu.

heat transfer   heat always flows toward a substance of lower temperature until the temperatures of the two substances equalize. It travels by one or more of three methods: conduction, convection, or radiation.

hertz   the units of measure of sound frequency, named for Heinrich H. Hertz. One hertz equals one cycle per second.

hip rafter   the diagonal rafter that forms a hip.

ice dam   condition formed at the lower roof edge by the thawing and refreezing of melted snow on the overhang. Can force water up and under shingles, causing leaks.

impact noise rating (INR) obsolete rating system for floor-ceiling construction in isolating impact noise. INR ratings can be converted to approximate IIC ratings by adding 51 points; however, a variation of 1 or 2 points might occur.

inside closure foam filler strip formed to fit the underside of corrugated sidewall and roof sheeting.

insulation, thermal any material that measurably retards heat transfer. There is wide variation in the insulating value of different materials. A material having a low density (weight/volume) usually is a good thermal insulator.

inverted track metal track installed upside-down over a first-floor wide-flange beam.

jackbeam a horizontal framing member used when an interior column has been eliminated.

jackrafter a short rafter, usually running between an eave and a hip rafter or between a ridge and a valley rafter.

jamb one of the finished upright sides of a door or window frame.

joist small beam that supports part of the floor, ceiling, or roof of a building.

joist hanger metal shapes formed for hanging on the main beam to provide support for the end of a joist.

lally column a circular steel pipe for supporting girders and beams.

ledger strip strip fastened to the bottom edge of a flush girder to help support the floor joists.

level true horizontal. Also a tool used to determine level.

lf linear feet, running feet.

limiting height maximum height for design and construction of a partition or wall without exceeding the structural capacity or allowable deflection under given design loads.

lintel horizontal member spanning an opening such as a window or door. Commonly referred to as a header.

live load that part of the total load on structural members that is not a permanent part of the structure. Can be variable, as in the case of loads contributed by the occupancy, and wind and snow loads.

load force provided by weight, external, or environmental sources such as wind, water, and temperature, or other sources of energy.

load combinations the value for live load in the same calculation is the larger of live load and snow load. This value is combined with dead load and collateral load.

loudness subjective response to sound pressure, but not linearly related thereto. A sound with twice the pressure is not twice as loud.

louver opening with slanted fins (to keep out rain and snow) used to ventilate attics, crawl spaces, and wall openings.

mansard roof a type of roof containing two sloping planes of different pitch on each of four sides. The lower plane has a much steeper pitch than the upper, often approaching vertical. Contains no gables.

mass property of a body that resists acceleration and produces the effect of inertia. The weight of a body is the result of the pull of gravity on the body's mass.

mastic roll caulking furnished to seal roof panels and for use with outside closure strips.

metal roll flashing used to weatherproof the junction between a high wall and low roof.

millwork includes all of the specialty wood items that are custom made in a cabinet shop and then installed. Generally furnished ready for final sanding and finishing. Can be purchased ready for pickup, delivered, or installed. Not to be confused with wooden trims, which are purchased in stock lengths, then cut and fitted to the house.

miter   joint formed by two pieces of material cut to meet at an angle.

modulus of elasticity (E)   ratio between stress and unit deformation, a measure of the stiffness of a material.

moment of inertia (I)   calculated numerical relationship (expressed in in.[4]) of the resistance to bending of a member, a function of the cross-sectional shape and size. A measure of the stiffness of a member based on its shape. Larger moments of inertia indicate greater resistance to bending for a given material.

mullion   vertical bar or division in a window frame separating two or more panes.

muntin   horizontal bar or division in a window frame separating multiple panes or lights.

N/A   not applicable.

nail pop   the protrusion of the nail usually attributed to the shrinkage or use of improperly cured wooden framing.

nominal   term indicating that the full measurement is not used; usually slightly less than the full net measurement, as with 2-×-4-inch studs that have an actual size when dry of $1\frac{1}{2} \times 3\frac{1}{2}$ inches.

nonbearing partition   a dividing wall that supports none of the structure above it.

nonveneer panel   any wood-based panel that does not contain veneer and carries an APA span rating, such as waferboard or oriented-strand board (OSB).

outside closure   foam filler strip formed to fit the top side of corrugated roof and sidewall sheeting.

overhang   the roof extension over a sidewall or endwall.

ozone resistance   a membrane's resistance to long-term exposure to ozone. Ozone rapidly degrades many rubber materials, especially if they are stressed.

panel   wood, glass, plastic, or other material set into a frame, such as in a door. Also a large, flat, rectangular building material such as plywood, hardboard, or drywall.

parapet wall   extension of an exterior wall above and/or through the roof surface.

partition   an interior dividing wall. Partitions might or might not be bearing.

penny (d)   a suffix designating the size of nails, such as 6d (penny) nail, originally indicating the price, in English pence, per 100 nails. Does not designate a constant length or size, and varies by type (i.e., common and box nails).

perlite board   heat and solvent resistant, fair R-value, low peel resistance, good fire resistance, and dimensional stability.

pilaster   projecting, square column or stiffener forming part of a wall.

pillar   column supporting a structure.

pitch of roof   slope of the surface, generally expressed in inches of vertical rise per 12 inches of horizontal distance, such as 4 in 12 or 4/12 pitch.

plastic laminate   generic term for commonly recognized trade names, such as Formica or Wilsonart countertops.

plate   top plate is the horizontal member fastened to the top of the studs or wall on which the rafters, joists, or trusses rest; sole plate is positioned at the bottom of the studs or wall.

platform   floor surface raised above the ground or floor level.

platform framing technique of framing where walls can be built and tilted up on a platform floor; in multistory construction are erected sequentially from one platform to another.

plenum a chamber in which the pressure of the air is higher (as in forced-air furnace systems) than that of the surrounding air. Frequently a description of the space above a suspended ceiling.

post-and-beam a method of construction that requires fewer, but heavier structural members than other methods. Also called plank-and-beam.

purlin horizontal member in a roof supporting common rafters, such as at the break in a gambrel roof. Also horizontal structural member perpendicular to main beams in a flat roof.

racking forcing out of plumb of structural components, usually by wind, seismic stress, or thermal expansion or contraction.

rafter that member forming the slanting frame of a roof or top chord of a truss. Also known as a hip, jack, or valley rafter depending on its location and use.

rafter tail that part of a rafter that extends beyond the wall plate: the overhang.

rake roof extension projecting over an endwall following the slope of the roof.

re-bar reinforcing steel used to increase the tensile strength of concrete. The size designation (e.g., #5) is the size in eighths of an inch. For example, #5 equals ⅝ inch diameter.

reglet a groove in a wall or other vertical surface adjoining a surface for the embedment of counterflashing.

ribs raised edges or folds that are formed in sheetmetal to provide stiffness.

ridge peak of a roof where the roof surfaces meet at an angle. Also can refer to the framing member that runs along the ridge and supports the rafters.

rim track metal track installed over the end of floor joists to close off the space between them.

rise measurement in height of an object; the amount it rises. The converse is fall.

riser vertical face of a step supporting the tread in a staircase.

roll flashing galvanized or aluminum flashing provided in rolls to be field-cut and formed to prevent water infiltration.

roll roofing coated felts, either smooth or mineral-surfaced.

rough framing structural elements of a building or the process of assembling elements to form a supporting structure where finish appearance is not critical.

R-value a measure of the resistance an insulating material offers to heat transfer. The higher the R-value, the more effective the insulation.

saddle the portion of a roof that connects two other roofs.

safing fire-stop material in the space between the floor slab and curtain wall in multistory construction.

self-drilling screws a fastener with a drilling point able to penetrate heavy-gauge metal.

self-tapping screws a sheetmetal screw used to join two pieces of light-gauge metal.

sheathing plywood, gypsum, wood fiber, expanded plastic, or composition boards encasing walls, ceilings, floors, and roofs of framed buildings. Can be structural or nonstructural, thermal-insulating or noninsulating, fire-resistant or combustible.

shed dormer a dormer projection with a single slope to the roof.

sheeting corrugated sheetmetal or exterior finishing material.

shingle mold wooden trim installed at the tip of the fascia board, directly under the leading edge of the roof shingles and behind the drip edge.

shipping list a list of part numbers and quantities of materials or components shipped with homes.

shoring temporary member placed to support part of a building during construction, repair, or alteration; also to support the walls of an excavation.

sill horizontal member at the bottom of door or window frames to provide support and closure.

slab a flat (although sometimes ribbed on the underside) reinforced concrete element of a building that provides the base for the floor or roofing materials.

soffit undersurface of a projection or opening; the bottom of a cornice between the fascia board and the outside of the building; the underside of a stair, floor, or header.

soil pipe a large pipe that carries liquid and solid wastes to a sewer or septic tank.

sound absorption conversion of acoustic or sound energy to another form of energy, usually heat.

sound insulation, isolation use of building materials or constructions that reduce or resist the transmission of sound.

sound transmission class (STC) single-number rating for evaluating the effectiveness of a construction in isolating audible airborne sound transmission across 16 frequencies. Higher numbers indicate more effectiveness.

spalling fragmenting or breaking up of the surface, due to freeze-thaw reaction within the wall.

span distance between supports, usually a beam or joist.

spandrel beam horizontal member, spanning between exterior columns, that supports the floor or roof.

spandrel wall exterior wall panel, usually between columns, that extends from the window opening on one floor to one on the next floor.

splice a joint at which two pieces are joined to each other. Also the action of connecting the two pieces.

spreader bar a device used to distribute weight equally while lifting widespan frames.

stirrup hanger to support the end of the joist at the beam.

stop strip fastened to the jambs and head of a door or window frame against which the door or window closes.

stress unit resistance of a body to an outside force that tends to deform the body by tension, compression, or shear.

strike the plate on a door frame that engages a latch or dead bolt.

stub column a short vertical structural member that bolts between two other members such as a rafter and ceiling beam.

stud vertical load-bearing or nonload-bearing framing member.

subfloor rough or structural floor placed directly on the floor joists or beams, to which the finished flooring is applied. As with resilient flooring, an underlayment might be required between subfloor and finished floor.

substrate underlayment material to which a finish is applied or by which it is supported.

tagline a rope tied to the base of a column to assist in controlling the frame action while it is in motion.

tail   the part of a rafter, if any, between the rafter plate and an overhanging eave.

tensile fatigue resistance   a membrane's ability to resist cyclic-induced internal and external tensile forces, particularly at joints in the substrate.

tensile strength   maximum tensile stress that can be developed in a given material under axial-tensile loading. Also the measure of a material's ability to withstand stretching.

tension   force that tends to pull the particles of a body apart.

thermal break   material used to prevent or reduce the direct transmission of heat or cold between two surfaces.

thermal conductivity (k)   heat energy in a Btu per hour transferred through a 1-inch-thick 1-square-foot area of homogeneous material per degrees F; temperature difference from surface to surface.

thermal resistance (R)   resistance of a material or assembly to the flow of heat. It is the reciprocal of the heat-transfer coefficient.

thermal shock   the stress producing phenomenon resulting from sudden temperature changes in a roof membrane.

threshold   raised member at the floor within the door jamb. Its purpose is to provide a divider between dissimilar flooring materials, or serve as a thermal, sound, or water barrier.

tin shingles (step flashing)   used to weatherproof the junction of a brick chimney or brick wall and roof.

tongue-and-groove joint   a joint where the projection or a tongue of one member engages the mating groove of the adjacent member to minimize relative deflection and air infiltration; widely used in sheathing, flooring, and paneling. Tongues can be in V-shaped, rounded, or square.

top track   horizontal member of stud wall that ties all studs together at the top.

tread   horizontal plane or surface of a stair step.

trim   any material, usually decorative, used to finish off corners between surfaces and around openings.

trimmer   double joists or rafters framing the opening of a stairway well, dormer opening, etc.

truss   open, lightweight framework of members, usually designed to replace a large beam where spans are great.

U factor   coefficient of heat transfer, U equals 1 divided by (hence, the reciprocal of) the total of the resistances of the various materials, air spaces, and surface air films in an assembly. *See Thermal Resistance.*

UL   Underwriters Laboratories, Inc.

UL label   label displayed on packaging to indicate the level of fire and/or wind resistance of asphalt roofing.

underlayment   a thin, smooth sheet material, usually hardboard or plywood, laid over subflooring as a smooth base for application of thin finished flooring materials.

valley   the intersection of two roof slopes.

valley flashing   used at all roof valleys to weatherproof the valley joint or roof shingles.

vapor retarder   material used to retard the flow of water vapor through walls and other spaces where this vapor might condense at a lower temperature.

veneer plaster   calcined gypsum plaster specially formulated to provide specific workability, strength, hardness, and abrasion-resistant characteristics when applied in thin coats over veneer-gypsum base or other approved base. The term thin-coat plaster is sometimes used in reference to veneer plaster.

vermiculite   an expanded aggregate used in lightweight insulating concrete.

wainscot   the surface material on the lower part of an interior wall, if different from the material on the upper part.

wallboard   a material for finishing the surfaces of interior walls and ceilings that is manufactured in large sheets out of gypsum, wood, or mineral fibers.

water absorption resistance   resistance to penetration of liquid water into the membrane. Absorbed water can alter the properties of the membrane.

water vapor permeability   the ability of a membrane to retard the flow of water vapor.

weatherability   the ability of the membrane to resist weathering, e.g., degradation due to sun, wind, rain, etc.

web   the center section of an I beam. *Compare flange.*

weep hole   small aperture at the base of an exterior wall cavity intended to drain trapped moisture.

window stool   Caution: The material take-off in this book does not make provisions for construction practices that encapsulate the entire window opening with wooden trims.

wood-fiber board   heat and solvent resistant, fair R-value, good peel resistance, poor fire and water resistance.

zoning   ordinances regulating the ways in which a property can be used in any given neighborhood. Zoning laws can limit where you can locate a structure. *See Building Codes.*

INDEX

Boldface number indicate illustrations.

ABOUT THE AUTHORS

Robert Scharff has been a regular contributor of articles to leading how-to and trade publications for almost 50 years. He wrote his first book, *Plywood Projects for the Home Craftsman*, for McGraw-Hill in 1954. Since that time he has written almost 300 books, most of them in the building and construction fields. In addition, he has served as a consultant for various building-materials manufacturers.

Walls & Ceilings Magazine is independently owned and published for contractors, suppliers, and distributors engaged in trades such as drywall, lath, plaster, stucco, ceiling systems, partitions, steel fireproofing, seamless flooring, poured roof decks, concrete pumping, and most importantly, residential steel construction. With a circulation of 40,000, the magazine has been known as the "voice of the industry" since 1938.

The Editors of **Walls & Ceilings**
are proud to bring you —
The Residential Steel Framing Handbook.

We invite you to continue to
educate yourself on all aspects involved in
the residential steel framing industry by
subscribing to **Walls & Ceilings** magazine .

Each month you'll find a wealth of
industry information on products,
literature, construction tips, industry news,
calendar events and much more.